Science Miracles
No Sticks or Snakes

First Edition
1420/2000

Science Miracles
No Sticks or Snakes

Dr. Adel M.A. Abbas
Anne P. Fretwell

amana publications
Beltsville, Maryland USA

© Copyrights 1420 AH / 2000 AC by
Dr. Adel M.A. Abbas
&
Anne P. Fretwell

Library of Congress Cataloging-in-Publication Data

Abbas, Adel M. A. (Adel Mohammed Ali), 1931-
 Science miracles : no sticks or snakes / Adel M.A. Abbas, Anne P. Fretwell.
   p. cm.
 English and Arabic.
 Includes bibliographical references.
 ISBN 0-915957-97-3
   1. Koran--Evidences, authority, etc. 2. Koran and science. I. Fretwell, Anne P. II. Title.

BP130.7 .A169 2000
297.1'2268--dc21

99-088707

Published by:
amana publications
10710 Tucker Street, Suite B
Beltsville, Maryland 20705-2223 USA
Tel: (301) 595-5777 • Fax: (301) 595-5888
Email: amana@igprinting

Table of Contents

Introduction   1

Chapter One

The Symbolic Letters   5
The Proportionate Distribution of the Symbolic Letters in the Qur'an (7), Explanation of the Letters (7), The Theme of the Following Chapters (10).

Chapter Two

Language and Linguistic Miracle   11
Literacy, Knowledge and Science (12), The Letter *Nūn* (13), Significance of Rhyme in the Qur'an (14), The Letter *Nūn* and the Beautiful Names of God (16).

Chapter Three

Receivers Set on the Right and Left   17
The Symbolic Letter *Qāf* (17), Possible Significance (17), The Universe and Its Clock (17), The Oort Cloud (19), Extension of Earth and the Lithosphere (19), Creation of Man (20), Possible Weight of Our Spirit (*Rūḥ*) (20), Nearer Than the Venus Return to the Heart (21), Two Receivers Set on the Right and Left (21), Provision and Rehydration (21), The Universe's Natural Phenomena (22), Life and Death and Supra-Sonic Waves (22), Science Beyond Time (23).

Chapter Four

Patience Is a Virtue   25
The Symbolic Letter *Ṣād* (25), God Provides Absolute Protection to His Message on Earth (26), The Lessons of Patience (26). *God's Patience with Ordinary People* (26), *God's Patience with His Prophets* (27), *God's Patience with the Devil* (27), *The Symbol of Patience* (28).

Chapter Five

The Five 'W's: Where, Why, Who, When and What 29
The Symbolic Letters *Ṭā Hā* (29), *Traditional Explanation* (29), *Computer Search* (29), *Interpretation in Relation to Other* Surah*s and the Names of God* (30), Explanation of *Ṭā Hā* in the Light of God's Names (31), The Guide (*Al Hādī*) (31), Prophet Moses — An Example of Guidance (31), *Where? and Why?* (31), *Who?* (32) The Technique the Prophet Moses Used to Spread His Message (32), Guidance Since Adam (33), The Objective of Guidance (33).

Chapter Six

Life, Death and Peace 35
High Mystical Value (35), The Symbolic Letters *Yā Sīn* (35), *Traditional Explanation of the Letters* (35), *Computer Search* (35), Explanation in the Light of the Names of God (35), *Power of Giving Life* (*Yuḥyī*) *and Causing Death* (*Yumīt*) (36), Peace (*Salām*) (36).

Chapter Seven

Imaging 37
The Symbolic Letters *Alif Lām Mīm Ṣād* (37), *Computer Search* (37) Creation and Imaging of Man (38), Magnetic Resonance Imaging (39), Mirror Image (40), Genetic Map (41), The Human Genome Project (41), Human Time Travel Transfer (42), Human Extension (43).

Chapter Eight

Parthenogenesis and Regeneration of Organs 45
The Symbolic Letters *Kāf Hā Yā 'Ayn Ṣād* (45), *Computer Search* (45), Explanation of the *Surah* in the Light of the Beautiful Names of God (46), Regeneration of Senile Organs (47), Reproduction by a Process Similar to Parthenogenesis (48), Similarity of Regeneration and Parthenogenesis (48) Guidance (50), God Does Not Need a Son (50).

Chapter Nine

The Equitable (*Al-Muqsiṭ*) 51
The Symbolic Letters: *Ṭā Sīn Mīm* and *Ṭā Sīn* (51), *Computer and Arabic Dictionary Search* (51), *Explanation of the Symbolic Letters* Ṭā Sīn Mīm *in the Light of the Beautiful Names of God* (51), Relation among the Three *Surah*s (52), Possible Relation between the Three *Surah*s and Other *Surah*s (53), Historical Signs of the Universe's Natural Phenomena (55), Signs and Science (56): *Social Sciences* (56), *Physical Sciences* (56), Science References — More Befitting Signs in Our Times than Sticks That Transform into Snakes (57), Planned Universe's Natural Phenomena (60).

Science Miracles: No Stick or Snakes vii

CHAPTER TEN

The Universe's Natural Phenomena   61

    The Flood of Noah: The Qur'anic Description (61), *Preparation for the Flood* (61), *Description of the Flood* (62), Possible Scientific Explanations (63), *Melting of the Ice Caps* (63), *Meteorite Impact* (65), *Tsunami or Seismic Sea Waves* (66), Meteor, Meteorites, Asteroids and Comets (68), The Blast That Turned Land Upside Down and the Rain with Labeled Stones from Hell (73), *The Qur'anic Description* (74) *Possible Scientific Explanation* (75), Screaming Cold Wind (77), *The Qur'anic Description* (77), *The Scientific Explanation* (78), Earthquakes (80), *The Qur'anic Description* (81), *Scientific Explanation* (82), Day of the Overshadowing Cloud (83) *The Qur'anic Description* (84), *Scientific Explanation* (84), Splitting the Water (85), *The Qur'anic Description* (86), Scientific Explanation (87), Man As Successor of God on Earth (88), Historical Universe Natural Phenomena, Science Miracles (89).

CHAPTER ELEVEN

Initiation of Signature   91

    Signs Are Miracles and/or Verses (92), Selective Saving from Mass Destruction (93), Scientific Allusions As Science Miracles (93), *Science Allusions Related to the Universe* (93), *Science Allusions Related to Earth* (94), *Science Allusions Related to Man* (95), The Universe's Natural Phenomena (95), Heavenly Books (95), God's Power (96).

CHAPTER TWELVE

The Merciful (*Al-Raḥmān*) Prevails throughout the Qur'an   97

    Scientific Allusions as Signs, Miracles (*Surah*s 40–46) (98), *Spirit from His Command* (98), *Physiology* (98), *Science Allusions Related to the Universe* (98), *Science Allusions Related to Earth* (98), *Science Allusions Related to Man* (99), *The Universe's Natural Phenomena* (99), Heavenly Books and God's Power (99), God Has No Son or Human Parts (99).

CHAPTER THIRTEEN

The Body of Knowledge and Advice   101

    Scientific Allusions (101): *Allusions Related to the Qur'an* (101), *Allusions Related to Man* (102), *Allusions Related to Public Health and Diagnosis of Pregnancy* (102), *Allusions Related to the Water Cycle* (102), *Allusions Related to the Universe* (103), *Allusions Related to the Fourth Dimension* (103), *Allusions Related to the Universe's Natural Phenomena* (103), God's Power (103).

Chapter Fourteen

The Light Circuit — 105

Man Is Driven by Energy (105): *Origin of the Energy* (105), *Extra Energy for Extra Duties* (106), The Weight of Our *Rūḥ* (106), The Dependance of Man on Electrochemical Reactions (106), The Light Circuit Path (110), *The Light Source of the Universe* (110), *The Light Path from Heaven to Earth* (110), The Origin of the Qur'an (111), The Light in the Qur'an (112), *Building Up of the Light Circuit in the Qur'an* (112), The Spread of Light into Man (118), Completion of the Light Circuit (120), Light Points Can Only Be Gained on Earth (122), The Symbolic Letters Are Initiators and Seal of the Light Circuit (122).

Chapter Fifteen

Heaven's Library — 123

Umm al-Kitāb (123), *Al-Lawḥ Al-Maḥfūẓ* or the Preserved Tablet (123), Books Related to the Universe and Earth (124), Books For Statistics (124), Books for Nations and Books for Individuals (124), The Size of These Books (125), The Symbolic Letters and Their Possible Role as Computer Markers (126).

Chapter Sixteen

Conclusion — 127

Bibliography — 137

Appendix

Rhyme or the *Qāfiya* at the End of the Verses of the Qur'an — 139
[Verse number and rhyming letter plus proportional distribution of rhyming letters for all *surah*s.]

INTRODUCTION

The Qur'an is a linguistic miracle by the standard of any language. It is neither prose nor poetry but a uniquely woven combination of both. The 6,236 verses (*ayāt*) that form its 114 chapters (*surah*s) are threaded together by loose rhymes into shorter or longer sequences. The rhythms of these sequences vary sensibly according to the subject matter. They swing from the steady march of the straight forward social and physical sciences to the majesty of Allah (God), the eminence of the Day of Judgement, the torments and terror of Hell, and the joys and delights of Paradise. Done in great splendor, with dramatic impact and moving beauty, the Qur'an is the only book in which over 50 percent of its verses rhyme with one concluding letter and almost 80 percent rhyme with three sounds. This fact has made even non-Arabic readers acknowledge its supremacy and inimitability.

The Qur'an contains miracles that are far beyond its inimitable language and which were meant to manifest millennia later. The people of Muhammad's time asked, "Why doesn't God send us miracles such as transforming a stick into a snake or an ordinary hand into one that is radiant as He did during the time of Musa (Moses)?" On one occasion, Allah replied (through the revelation of the Qur'an) by making a "scientific" allusion: "All creatures which move with a sound, on or in the earth, and birds that fly with wings, are nations like ourselves." The significance of this statement still challenges us. We scarcely understand how whales and dolphins call one another and have only recently appreciated the body language of bees. Allah knew that Muhammad's nation — those born from the time that the Prophet Muhammad was delivering the message until the Day of Doom — would come to live at a time when scientific facts relating to the Universe and man would be revealed. Hence, Qur'anic miracles are given as scientific allusions which, when recognized, acknowledge Allah as the Creator.

Some of the scientific allusions in the Qu'ran have recently been recognized; for example, the difference in time between the Earth and the Universe. This is illustrated by several clocks: the day and the night, the lunar month and year, the solar year for the Earth and planets, and the fourth dimension for objects in space. Time with the Creator is eternal. The heavens and earth were indistinguishable as smoke, then, through a long process, the Earth became disinct and separate. Creation is a chain of events that began with a "big bang" in which all the energy of the Universe was produced. It started with the formation of particles, then atoms of hydrogen which formed smoke, nebulae, then stars like our sun, and then planets like our Earth. The high energy which began the creation must have had an origin. In the Qur'an, God says that the sun and the moon "swim in their own orbits." We also know the sun has a life span and that its light will eventually be exhausted. Many scientific allusions in the Qur'an concerning the Universe have been found but many are as yet to be explained.

Some of the scientific allusions in the Qur'an concerning the Earth have now been identified. He extends and reduces the Earth in areas that are unseen by man. The process takes place under the oceans. He extends the Earth at the oceanic ridges and reduces it at the Pacific trenches. We have now confirmed the shape and formation of the Earth. Its layers include the lithosphere (*rawasi*), which is there "lest the earth cave in from under us," and which is used for extension and reduction of the Earth. Mountains act as pegs to stop the lithosphere from going inside the Earth. We now find that all mountains have roots that extend into the earth to a greater depth than its visible part climbs above the earth's surface. Many Qur'anic signs relating to the earth are now familiar, including the water cycle, the estuarine cycle, runaway mountains, and earthquakes. While many allusions have yet to be understood.

Many scientific references in the Qur'an concerning man are now recognized but many are still obscure. We have no instruments sufficiently sensitive to measure the difference in the weight of the body before and after death, thus sensing the weight of the soul — if it has weight.

In the Qur'an, Allah also gives us examples of the universe's natural phenomena that He used in the past to destroy nations. Many examples are given: the flood of Noah in which the sea level rises considerably above its normal level producing mountainous waves accompanied by suprasonic sound waves, hurricane winds, or bolts of lightening, as well as fire falling to Earth from outer space, sulphuric acid clouds, and brimstone rain. Such historical reports were warnings to Muhammad's nation. Fortunately, many countries are now aware of natural phenomena such as meteorites and asteroids that can cause destruction and fires on Earth. Some countries have observation programs and have developed missiles to either stop the meteorites and asteroids or deter them from reaching the Earth or its atmosphere. Man, being God's steward on Earth, has a duty to protect his place of residence.

In addition to its miraculous descriptions of physical events, the Qur'an also contains miraculous symbolic letters that appear at the beginning of twenty-nine of its *surahs*. For fourteen hundred years, they have challenged man to explain them. God has asked us to reflect on His creation and to work hard to understand His verses and His signs. This book attempts to explain these symbolic letters in the light of the beautiful names of God, which not only control everything in the universe but also control the system of guidance He has provided for man. We will show that these letters could be symbolic of a light circuit as modeled in the Qur'an, revealed from God through the prophets through their holy books and through the Qur'an. The system of lights is very complex as it is channelled through ninety-nine adjectives of the beautiful names of God. The primary part of the circuit is channelled through the symbolic letters inscribing the name of God, *Al-Rahman* (the Merciful), through and across the Qur'an. Most of the name, except the letter *nūn*, is inscribed at the beginning of the *surahs* in the middle of the Qur'an. It is complemented by a base of 3,123 *nūn*s rhyming 50.08 percent of the verses. A radar graph shows that *Al-Rahman* maintains a straight and unique path. Light from the Qur'an and from His guidance is returned to Him through man. This is effected by man when he takes part in a process of learning and acting on what he has learned, thus creating light impulses leading to a scoring system. This probably happens in our brain, programmed by God, in areas as yet unknown. A printout of all these light impulses will perhaps form the book that will be presented to each of us on the Day of Judgment. For those lucky people with a high score, light will shine around them qualifying them for reward.

One of the most amazing facts in the Qur'an is the reference to an endless number of books. Not only are there books for each human being but also for every creature that makes a sound on or in the Earth. There are books for nations, books for the universe, and books for each of its contents. How are all these books and their detailed contents, as related to themselves and to each other, going to be collated, classified, filed and retrieved? The size and complexity of such a "library" is beyond comprehension. This book suggests that the symbolic letters initiate light circuit signs, thus acting like computer markers. Some of these symbolic letters are arranged so that they appear to write, in light, the name of God, *Al-Rahman*, to shine across the Qur'an as a way of distinguishing it from all other books collated in *umm al-kitāb* (the Mother Book).

Chapter One

The Symbolic Letters

On twenty-nine occasions Allah begins a *surah* of the Qur'an with one or more symbolic letters. These letters are sometimes at the beginning of the first verse and at other occasions they stand alone. They are referred to as *al-muqāṭṭa'āt*, an Arabic term indicating that they are parts of words. Hence, they are called "abbreviated" or "symbolic" letters. This nomenclature is assumed, but as yet nobody knows what these letters stand for. They are sometimes also called "Openers of Surahs."

Collectively, they represent fourteen of the twenty-nine letters of the Arabic alphabet. The fourteen letters are:

ا ص ك ه ح ط ل ى ر ع م س ق ن

The entire Arabic alphabet contains the following letters:

ا ب ت ث ج ح خ د ذ ر ز س ش ص ض ط ظ ع غ ف ق ك ل م ن ه و ي ة ء

Table 1. shows the symbolic letters that stand alone according to their order of revelation.

Table 1: Symbolic Letters in the Qur'an.

Symbolic Letter(s)	Cronological Order	Name of Surah	Order in Qur'an	Number of Verses
ن	2 (BH)	Qalam	68	52
ق	34 (BH)	Qāf	50	45
ص	38 (BH)	Ṣād	38	88
المص	39 (BH)	A'rāf	7	206
يس	41 (BH)	Yasīn	36	83
كهيعص	44 (BH)	Maryam	19	98
طه	45 (BH)	Tā Hā	20	135

طسم	47 (BH)	Shuʻarā	26	227
طس	48 (BH)	Naml	27	93
طسم	49 (BH)	Qaṣaṣ	28	88
الر	51 (BH)	Yūnus	10	111
الر	52 (BH)	Hūd	11	123
الر	53 (BH)	Yūsuf	12	111
الر	54 (BH)	Hijr	15	99
الم	57 (BH)	Lukmān	31	34
حم	60 (BH)	Mu'min	40	85
حم	61 (BH)	Fussilat	41	54
حم عسق	62 (BH)	Shūrā	42	53
حم	63 (BH)	Zukhruf	43	89
حم	64 (BH)	Dukhān	44	59
حم	65 (BH)	Jāthiyah	45	37
حم	66 (BH)	Ahqāf	46	35
الر	72 (BH)	Ibrāhīm	14	52
الم	75 (BH)	Sajdah	32	30
الم	84 (BH)	Rūm	30	60
الم	85 (BH)	ʻAnkabūt	29	69
الم	87 (AH)	Baqarah	2	286
الم	89 (AH)	Al-ʻImrān	3	200
المر	96 (AH)	Ra'ad	13	43

When all of the symbolic letters in the twenty-nine *surah*s that contain them are added together they come to eighty letters. Their distribution is displayed in Table 2.

Table 2: Total Number of Each of the Symbolic Letters.

ن	ه	ي	ص	ك	ع	س	ق	ح	م	ل	ا	ط	ر
1	2	2	3	1	2	5	2	8	18	13	13	4	6

The Proportionate Distribution of the Symbolic Letters in the Qur'an

The symbolic letters represent a small portion of the letters that make up the Qur'an (see Table 3).

Table 3: Proportionate Distribution of Symbolic Letters in the Qur'an.

ه	ك	ص	ا
4.49%	3.174%	.626%	13.168%
ى	ل	ط	ح
6.652%	11.541%	.385%	1.252%
	م	ع	ر
	8.087%	2.844%	3.75%
	ن	ق	س
	8.244	2.127%	1.818%

Explanation of the Letters

Most Qur'anic exegetes agree that the letters are presented by Allah at the beginning of twenty-nine *surah*s as a challenge to the Arab speaking people. The Qur'an descended among the Arabs in their language as a miraculous book of beautiful, closely woven Arabic. On several occasions, Allah challenges the Arabs to produce one *surah* like those that are in the Qur'an (2:23; 10:38; 17:88).

Like millions before us and millions who will come after us, we wonder what the symbolic letters stand for. Are they the initials of the beautiful names of Allah? Or are they part of them? Do they refer to the verses that come after them? Do they indicate places, events, other *surahs*? These are questions that remain to be answered. No human being can claim knowledge of what they mean. In fact, nobody will ever know what they are there for; however, we have been asked not only to reflect on Allah's creation (2:164) but also to consider and study His verses and signs (38:29). Allah has mentioned that some of the verses in the Qur'an are fundamental and are, in fact, *umm al-kitab* (literally, the mother of the book; meaning, the core of the Qur'an), containing the central message for man. Other verses are allegorical. Some people in whose heart is perversity follow the part that is allegorical, seeking discord (3:7). Our sincere intention is to seek the truth using today's scientific knowledge to pursue the meaning of the symbolic letters.

Behold! In the creation of the heavens and the earth; in the alternation of the night and the day; in the sailing of the ships through the ocean for the profit of mankind; in the rain which God sends down from the skies, and the life which He gives therewith to an earth that is dead; in the beasts of all kinds that He scatters through the earth; in the change of the winds, and the clouds which they trail like their slaves between the sky and the earth; (here) indeed are signs for a people that are wise. (2:164)

إنَّ فِي خَلْقِ السَّمَوَاتِ وَالأَرضِ وَاخْتِلَافِ اللَّيْـلِ وَالنَّـهَارِ وَالْفُلْكِ الَّتِي تَجْرِي فِي الْبَحْرِ بِمَا يَنْفَعُ النَّاسَ وَمَا أَنْزَلَ اللَّهُ مِنَ السَّمَاءِ مِنْ مَاءٍ فَأَحْيَا بِهِ الأرضَ بَعْدَ مَوْتِهَا وَبَثَّ فِيـهَا مِنْ كُلِّ دَابَّةٍ وَتَصْرِيفِ الرِّيَاحِ وَالسَّحَابِ الْمُسَـخَّرِ بَيْـنَ السَّمَاءِ وَالأرضِ لَآيَاتٍ لِقَوْمٍ يَعْقِلُونَ (١٦٤)

Surat al-Baqarah

Here is a book which We have sent down unto thee, full of blessings, that they may meditate on its signs, and that men of understanding may receive admonition. (38:29)

كِتَابٌ أَنزَلْنَاهُ إِلَيْكَ مُبَارَكٌ لِيَدَّبَّرُوا آيَاتِهِ وَلِيَتَذَكَّرَ أُولُوا الأَلْبَابِ (٢٩)

Surah Ṣād

He it is who has sent down to thee the Book: In it are verses basic or fundamental (of established meaning); they are the foundation of the Book; others are allegorical. But those in whose hearts is perversity follow the part thereof that is allegorical, seeking discord, and searching for its hidden meanings execpt God. And those who are firmly grounded in knowledge say: "We believe in the Book; the whole of it is from our Lord: and none will grasp the Message except men of understanding. (3:7)

هُوَ الَّذِي أَنْزَلَ عَلَيْكَ الْكِتَابَ مِنْهُ آيَاتٌ مُحْكَمَاتٌ هُـنَّ أُمُّ الْكِتَابِ وَأُخَرُ مُتَشَابِهَاتٌ فَأَمَّا الَّذِينَ فِـي قُلُوبِـهِمْ زَيْـغٌ فَيَتَّبِعُونَ مَا تَشَابَهَ مِنْهُ ابْتِغَاءَ الْفِتْنَةِ وَابْتِغَاءَ تَأْوِيلِهِ وَمَا يَعْلَـمُ تَأْوِيلَهُ إِلاَّ اللَّهُ وَالرَّاسِخُونَ فِي الْعِلْمِ يَقُولُونَ آمَنَّا بِهِ كُـلٌّ مِنْ عِنْدِ رَبِّنَا وَمَا يَذَّكَّرُ إِلاَّ أُولُو الأَلْبَابِ (٧)

Surat Al 'Imrān

Science Miracles: No Stick or Snakes 9

Searching for the truth, the safest ground for our explanation is the beautiful names of God. Allah asks us to use them in our prayers (7:180; 17:110; 20:8; 59:24). We approached this theme of research systematically.

If the fourteen letters are to be used once only — even when added together as eighty letters — very few names can be formulated and many letters will remain unused. If, however, the fourteen letters are considered constituents that may be repeated, several names can be formulated. Examples are shown below in Table 4.

Table 4: Possible Names.

الرحمن Al-Rahmān The Merciful	الرحيم Al-Rahīm The Compassionate	الملك Al-Malik The Sovereign	السلام Al-Salām The Peace	المهيمن Al-Muhaymin The Protector
القهار Al-Qahār The Supreme	العليم Al-'Alīm The All-Knowing	السميع Al-Samī' The All-Hearing	الحكم Al-Hakam The Judge of All	الحليم Al-Halīm The Forbearer
العلى Al-'Alī The High	الكريم Al-Karīm The Most Generous	الحكيم Al-Hakīm The Wise	الحق Al-Haqq The True	الحميد Al-Hamīd The Praiseworthy
المحصى Al-Muhṣī The Knower of each separate thing	المحيى Al-Muhyī The Life Giver	الحى Al-Hayy The Living	المقسط Al-Muqsit The Equitable	المانع Al-Māni' The Withholder

The 99 names of Allah contain all the twenty-nine letters of the alphabet. So it is impossible that the fourteen letters could formulate all the 99 names. A list of the names are shown in Table 5.

Table 5: The 99 Beautiful Names of God.

القدوس Al-Quddus The Holy One	الملك Al-Malik The Sovereign	الرحيم Al-Rahim The Compasionate	الرحمن Al-Rahman The Merciful	الله Allah God
الجبار Al-Jabar The All-Compelling	العزيز Al-Aziz The Mighty	المهيمن Al-Muhaymin The Protector	المؤمن Al-Mu'min The Faithful	السلام Al-Salam The Peace
الغفار Al-Ghafar The Forgiver	المصور Al-Musawwir The Imager	البارئ Al-Bari' The Maker	الخالق Al-Khaliq The Creater	المتكبر Al-Mutakabbir The Exalted
العليم Al-'Alim The All Knowing	الفتاح Al-Fattah The Opener	الرزاق Al-Razzaq The Provider	الوهاب Al-Wahhab The Bestower	القهار Al-Qahar The Supreme
المعز Al-Mu'izz The Bestower of Honor	الرافع Al-Rafi' The Exalter	الخافض Al-Khafid The Abaser	الباسط Al-Basit The Enlarger	القابض Al-Qabid The Closer
العدل Al-'Adl The Just	الحكم Al-Hakam The Arbitrator	البصير Al-Basir The All-Seeing	السميع Al-Sami' The All-Hearing	المذل Al-Mudhill The Subduer
الغفور Al-Ghafur The Pardoner	العظيم Al-'Azim The Great	الحليم Al-Halim The Forbearer	الخبير Al-Khabir The Well Aware	اللطيف Al-Latif The Benignant

الشكور	العلى	الكبير	الحفيظ	المقيت
Al-Shakūr	Al-'Aliy	Al-Kabīr	Al-Ḥafiẓ	Al-Muqīt
The Grateful	The High	The Greaty	The Guardian	The Maintainer
الحسيب	الجليل	الكريم	الرقيب	المجيب
Al-Ḥasīb	Al-Jalīl	Al-Karīm	Al-Raqīb	Al-Mujīb
The Reckoner	The Majestic	The Most Generous	The Watchful	The Answerer of Prayers
الواسع	الحكيم	الودود	المجيد	الباعث
Al-Wāsi'	Al-Ḥakīm	Al-Wadūd	Majīd	Al-Bā'ith
The Vast	The Wise	The Loving	The Glorious	The Raiser of the dead
الشهيد	الحق	الوكيل	القوى	المتين
Al-Shahīd	Al-Haqq	Al-Wakīl	Al-Qawi	Al-Matīn
The Witnesser	The True	The Guardian	The Strong	The Firm
الولى	الحميد	المحصى	المبدئ	المعيد
Al-Waliy	Al-Ḥamīd	Al-Muḥsi	Al-Mubi'	Al-Mu'id
The Friend	The Praiseworthy	The Knower of each separate thing	The Beginner	The Repeater
المحيى	المميت	الحى	القيوم	الواجد
Al-Muḥyī	Al-Mumīt	Al-Ḥayy	Al-Qayyūm	Al-Wājid
The Life Giver	The Slayer	The Living	The Self-Subsisting	The Resourceful
الماجد	الواحد	الصمد	القادر	المقتدر
Al-Mājid	Al-Wāḥid	Al-Ṣamad	Al-Qādir	Al-Muqtadir
The Origin of Glory	The One	The Eternal	The All-Powerful	The All-Determiner
المقدم	المؤخر	الأول	الآخر	الظاهر
Al-Muqaddim	Al-Mu'akhkhir	Al-Awwal	Al-Ākhir	Al-Ẓāhir
The Promoter	The Postponer	The First	The Last	The Manifest
الباطن	الوالى	المتعالى	البر	التواب
Al-Bāṭin	Al-Wāli	Al-Muta'ali	Al-Barr	Al-Tawwāb
The Hidden	The All-Governing	The Exalted	The Good	The Acceptor of Repentance
المنتقم	العفو	الرؤوف	مالك الملك	ذو الجلال والإكرام
Al-Muntaqim	Al-'Afūw	Al-Ra'ūf	Mālik al-Mulk	Dhu Al-Jalali wa Al-Ikrām
The Avenger	The Effacer of Sins	The Most Kind	The Lord of Sovereignty	The Lord of Majesty and Generosity
المقسط	الجامع	الغنى	المغنى	المانع
Al-Muqsiṭ	Al-Jāmi'	Al-Ghani	Al-Mughnī	Al-Māni'
The Equitable	The Gatherer	The Self-Sufficient	The Enricher	The Withholder
الضار	النافع	النور	الهادى	البديع
Al-Ḍārr	Al-Nāfi'	Al-Nūr	Al-Hādi	Al-Badī'
The Hurtful	The Benefiter	The Light	The Guide	The Originator
الباقى	الوارث	الرشيد	الصبور	
Al-Bāqī	Al-Wārith	Al-Rashīd	Al-Ṣabūr	
The Everlasting	The Inheritor	The Right in Guidance	The Patient	

The Theme of the Following Chapters

It is our intention to use the beautiful names of God to be our safe guide in explaining the symbolic letters. Sometimes they appear to be part of one of the names of God, but when this is not obviously possible, they will be considered as initials of His beautiful names. The proposed explanation will be made in the context of the meaning of the *surah* in which they occur. Sometimes, when the symbolic letters occur in more than one place they will be considered as a group. They will also be correlated to other *surah*s when they seem to be related. As seen earlier, we chose to present the symbolic letters in cronological order of revelation so that we do not miss possible meanings that could be relevant to the order of descent of the Qur'an.

Chapter Two

Language and Linguistic Miracle

According to Muslim scholars, the Qur'an should not be translated. It is a language and linguistic miracle, and to be appreciated it should remain in Arabic. The Qur'an was first translated into Latin in 1143 C.E. It first appeared in English (from the Latin) in 1657, followed by the English translation directly from the Arabic by Sale in 1734. More recent translations were made by Rodwell (1861), Palmer (1880), and Pickthall (1930) were either termed "interpretation" (Arberry 1983), or "meanings" of the Qur'an. Most recently translations have been produced by eminent institutions in Saudi Arabia, Egypt, and other countries.

No translation can possibly be successful, even if it attempts to produce the meanings in a style that echoes the sublime rhetoric of the Arabic language of the Qur'an. We can understand the attitude of the strict scholars, when we witness the intricate and richly varied rhythms of the Qur'an — apart from the message itself — thus making it the premiere linguistic masterpiece of mankind. In the history of Islam, many people have been moved to join the faith by just listening to the Qur'an, the very sounds of which moved them to tears and ecstasy.

The Qur'an is neither prose nor poetry but a linguistic marvel uniquely woven in a language structured to combine both. Its verses are threaded together by loose rhymes into shorter or longer sequences within *surah*s. The rhythms of these sequences vary sensibly according to the subject matter, swinging from the steady march of straight forward narrative or articulation of the Majesty of God, to the immanence of the Day of Judgment, the torments and terrors of Hell and the joys and delights of paradise. All of this is done in great splendor, drama, and beauty. For this reason Allah calls it the Glorious Qur'an.

Literacy, Knowledge and Science

The first few verses that descended from God to man were mainly concerned with preparing Muhammad and his nation for the progress of knowledge and science during their time. This could only be affected through reading and writing. It is worth mentioning that the term "Muhammad's nation" refers to all people who would be born after Muhammad's revelation until the Day of Resurrection. God asked mankind to read in His name, the Creator, who created him from blastocysts (Abbas 1997). He asked him to write with the pen and grasp all the knowledge concealed from him (96:1–5). This great advice was conveyed through the Angel Gibril to the Prophet, who could neither read nor write (29:48), and was then conveyed to an illiterate people, most of whom simply thought he was mad (68:51). Appropriately, the next revelation (*Al-Qalam*/The Pen), assured them that he was not mad and that the Qur'an is an Arabic language miracle and not the work of a man (68:2,5).

Read in the name of thy Lord who created. Created man out of a (mere) clot of congealed blood (blastocyst). Proclaim! And thy Lord is Most-Bountiful. He who taught (the use of) the Pen. Taught man that which he knew not. (96:1–5)	اقْرَأْ بِاسْمِ رَبِّكَ الَّذِي خَلَقَ(١)خَلَقَ الإنْسَانَ مِنْ عَلَقٍ(٢)اقْرَأْ وَرَبُّكَ الأكْرَمُ(٣)الَّذِي عَلَّمَ بِالْقَلَمِ(٤)عَلَّمَ الإنْسَانَ مَا لَمْ يَعْلَمْ (٥) Surat al-'Alaq
And thou wast not (able) to recite a book before this (book came). Nor art thou (able) to transcribe it with thy right hand: In that came, indeed, would the talkers of vanities have doubted. (29:48)	وَمَا كُنْتَ تَتْلُو مِنْ قَبْلِهِ مِنْ كِتَابٍ وَلاَ تَخُطُّهُ بِيَمِينِكَ إِذًا لاَرْتَابَ الْمُبْطِلُونَ (٤٨) Surat al-'Ankabūt

The Letter *Nūn*

The first verse of *Surah al-Qalam* is a vow invoking the letter *nūn* and the might of the pen. It states: *"Nūn and the pen, and what it can write in lines and lines."* For the benefit of the non-Arabic reader the Arabic verse is transliterated. It reads: "Nūn, wa al-qala<u>m</u> wa ma yasturū<u>n</u>." The verse sets the tone for the whole rhyme scheme of the Qur'an. It rhymes with two sounds effected by the two underlined letters. These are the letters *nūn*, used to start and finish the verse, and the *mīm*, which is the last letter of *qalam* (pen).

Nūn, by the Pen and by what it writes in lines and lines. (68:1)   ن وَالْقَلَمِ وَمَا يَسْطُرُونَ (١)

Surat al-Qalam

There is evidence that this verse refers to the Qur'an itself (52:2). The *surah* continues to rhyme through most of its verses (88.8%) with the letter *nūn*. On occasions, for emphasis or variation, the letter *mīm* is used for rhyme (19.2%).

In a *surah* of fifty-two verses this does not seem to be miraculous; however, when the entire rhyming scheme of the Qur'an is analyzed in detail (see Appendix), it becomes obvious that the rhyme distribution constitutes one of the main linguistic miracles in the Qur'an. In over fifty percent (50.08%) of the 6,236 verses, rhyme is made with the letter *nūn*. As far as we are aware, no literary work of comparable size has a rhyming scheme with one sound or letter in more than half of its text. This applies not only to the Arabic language but also to all languages of mankind.

Rhyme analysis of the Qur'an shows that almost 80 percent of it rhymes with three concluding sounds (n, m, a) produced by the four letters *alif, mīm, yā, and nūn* (ا/ى م ن) (see Table 6).

One can understand how a poem of 200 to 300 lines might rhyme with two or three sounds and then be termed a "masterpiece." Usually the work deals with an emotional or descriptive subject in which the rhyme adds a musical element. The Qur'an is a large book whose subject is religion, guidance, and belief. It also covers many subjects that are included within criminal and civil law, psychology, and other social sciences. It mentions many subjects that deal with physical science relating to the universe, earth, relativity of time, and creation of man. It is, indeed, miraculous to cover these subjects while maintaining a rhyme based on so few sounds. The Arabs who mastered this language found it inimitable. Readers, regardless of their mother tongue, have agreed that the style is virtually impossible to imitate. Those who have tried have failed.

Table 6 shows the proportionate distribution of the four letters (three sounds) which formulate 79.92 percent of the rhyme of the Qur'an.

Table 6: The Four Most Common Rhyming Sounds.

Letter	ا	ى	م	ن	Totals
Sounds	a	a	m	n	
No. of Verses	949	246	666	3123	4984
Percent	15.22	3.94	10.68	50.08	79.92

Significance of Rhyme in the Qur'an

Analysis shows various patterns of rhyme in the Qur'an. The *nūn* appears in 72 of the 114 *surahs* (63.1%) as a final letter of a verse. Table 7 shows the proportional distribution of the verses ending with *nūn* in the *surahs* starting with symbolic letters. Later, we will show that its prevalence within these *surahs* points to a possible relationship with its content.

Table 7: Proportionate Distribution of Verses Ending with Nūn *in Surahs That Begin with Symbolic Letters.*

Surah	Name of Surah	Symbolic Letters	*Nūn*s	Surah	Name of Surah	Symbolic Letters	*Nūn*s
2	Baqarah	الم	193 (67.8%)	30	Rūm	الم	54 (90%)
3	Al Imrān	الم	121 (60.5%)	31	Lukmān	الم	7 (20.6%)
7	A'rāf	المص	193 (93.7%)	32	Sajdah	الم	27 (90%)
10	Yūnus	الر	98 (89.9%)	36	Yasin	يس	71 (85.5%)
11	Hūd	الر	56 (45.5%)	38	Ṣād	ص	18 (20.5%)
12	Yūsuf	الر	93 (83.8%)	40	Ghāfir	حم	32 (37.7%)
13	Ra'd	المر	5 (11.6%)	41	Fussilat	حم	30 (55.6%)
14	Ibrāhīm	الر	6 (11.5%)	42	Shūrā	حم عسق	6 (11.3%)
15	Hijr	الر	81 (81.8%)	43	Zukhruf	حم	78 (87.6%)

Surah	Name of Surah	Symbolic Letters	Nūns	Surah	Name of Surah	Symbolic Letters	Nūns
19	Maryam	كهيعص	5 (5.1%)	44	Dukhān	حم	44 (74.6%)
20	Ṭā Hā	طه	0 (0%)	45	Jāthiya	حم	30 (81.1%)
26	Shuʻarā	طسم	192 (84.6%)	46	Aḥqāf	حم	26 (74.3%)
27	Naml	طسم	84 (90.3%)	50	Qāf	حم	0 (0%)
28	Qaṣaṣ	طسم	81 (92.1%)	68	Qalam	ن	42 (88.8%)
29	ʻAnkabūt	الم	59 (85.5%)				

The letter *mīm* rhymes 10.68 percent of the verses of the Qur'an. As seen earlier, *mīm* is the terminal letter of *qalam* (pen). This rhyme is present in 71 percent of the *surah*s. Unlike the *nūn* rhyme, the *mīm* rhyme does not predominate, except in *surah* 47 titled "Muhammad," the name of the Prophet which starts with the letter *mīm*. The rhyme scheme of the *surah* is not by chance. Perhaps Allah uses it to honor the Prophet.

In another 19.16 percent of the Qur'an, the verses rhyme with the "aa" sound which is produced by either *alif* or *yā*. Both letters are pronounced as "aa" at the end of a sentence, in special conditions of conjugation. It is interesting to note that *surah*s ending with the "aa" sound seem to have common themes: some deal with women (see *surah*s 4, 19, and 65); others deal with miraculous events, God's powers, and His description of the physical world (see *surah*s 17, 18, 25, 48, 71, 72, 73, 76, 78, and 91).

Table 8 shows that 96.74 percent of the verses rhyme with only eight sounds. These are made by nine of the twenty-nine letters of the Arabic alphabet.

Table 8: Proportionate Distribution of Rhyme Sounds in All Verses of the Qur'an.

Arabic Letter	Nūn	Mīm	Alif	Yā	Rā	Dāl	Bā	Tā Marbuta	Hā	Lām	Other Letters	Total
	ن	م	ا	ى	ر	د	ب	ة	ه	ل		
Sound	n	m	aa	aa	r	d	b	t	h	l		
No. of Verses	3123	663	949	246	453	198	160	122	49	67	203	6236
Percent	50.08	10.68	15.22	3.94	7.26	3.18	2.57	1.96	0.79	1.07	3.25	100

This chapter illustrates how the Qur'an is inimitable in sounds and presentation. Rhyme is just one of the many linguistic dimensions in the Qur'an that puzzle the Arabic-speaking peoples. As shown earlier, the rhyme is not randomly spread, but is placed with calculated precision to produce the correct impact to deliver the meaning. In addition, *surah*s seem to be correlated by subject, theme, and style. An analytical study of this complex interrelatedness is beyond the scope of this book; however, the total study and the statistical analysis presented in the appendix are available on disk and can be requested from the authors. Serious scholars may wish to study in depth the different aspects of this magnificent, intricate network of rhyme and beauty.

The Letter *Nūn* and the Beautiful Names of God

The letter *nūn* begins two names of God — *Al-Nur* (Light) and *Al-Nāfi'a* (Beneficial) — which when combined together is "Beneficial Light" (النور النافع); and ends two other names of God — *Al-Rahman* (The Merciful) and *Al-Bātin* (The Innermost) — when combined together is "The Merciful Innermost" (الرحمن الباطن). If one replaces the first two names of God with the *nūn* present in the first verse of *surah* 68, it reads as follows: "Beneficial light is what the pen can produce in continuous lines." This is the first verse and is the subject of this chapter.

People live in darkness until they can read and write. Emphasis of this fact was made in the Qur'an many centuries ago and lead to the rapid progress of mankind.

CHAPTER THREE

RECEIVERS SET ON THE RIGHT AND LEFT

The Symbolic Letter *Qāf* (ق)

The letter *qāf* begins the first verse of *surah* 50 (34 cronologically) which takes the name of the letter itself. God uses the letter and the glorious Qur'an to adjure. The Qur'an is described as glorious, which is one of the names of God (*Al-Majid* مجيد).

Possible Significance

Unlike the letter *nūn* discussed in the previous chapter, which forms most of the rhyme of *Surah al-Qalam* and over half the rhyme of the Qur'an, the letter *qāf* does not appear at all as a rhyme in the *surah* which takes its name. In fact, the rhyme in *Surah Qāf* is completely different (see Appendix) — 60 percent is made by the letter *dāl*. If one reads the *surah*, the letter *qāf* stands out marking the words of 34 out of 45 verses (75.55%). Surprisingly, all the *qāf*-marked verses stand out as scientific allusions, some of which are very challenging. The science references would certainly seem mystical to a primitive society. Their explanation is very difficult and their classification is cumbersome. We found it much easier to classify them according to 16 names of God which contain the letter *qāf*.

The Universe and Its Clock

God created the heavens and the earth and what is between them in six days of the universal clock. It caused Him no fatigue (50:6,38). He who maintains His creation (i.e., *Al-Muqīt* المقيت) is Self-Subsisting (*Al-*

Qayyum القيوم). He ornamented the lower heaven with the Sun, the Moon, the planets, and the stars. Without this heaven, life on Earth would be impossible. No obvious openings can be seen in the sky. God has informed us in the Qur'an that there are at least four "clocks" that we should be aware of: the variation of the day and night (31:29); the lunar month (36:39); the twelve months in the year; and the solar system. The Sun "swims" in a special orbit (36:40) according to a precise span of life (13:2). It rises and sets on Earth at different points (70:40). It is shielded from sight as it sets (38:32). All this produces the solar year. The planets of the solar system have always been known and are referred to in the Qur'an as *kawakib*. Each has a solar year of a different length. Their solar year could be as short as 86.9793 days, as on Mercury which is the nearest to the sun, or as long as 90777.61 days, as on Pluto, the farthest planet from the Sun. The length of their seasons and days are also different (see Table 9). We now know that our sun is one of millions. Each sun has its orbit and some have planets. They exist in a four dimensional continuum called space-time. All this shows The Power *(Al-Qawwī* القوى) of The Creator *(Al-Khāliq* الخالق) and how He spreads His Authority *(Al-Qadr* القادر) through His kingdom.

Table 9: The Solar System.

Planet	Distance from Sun (km)	Length of Sidereal Day (d h m)			Length of Year (Solar Days)
Sun	—	25	09	07	—
Mercury	57909100	58	15	30	87.9693
Venus	108208900	243	00	14*	224.7008
Earth	149597900		23	56	365.2564
Mars	227940500		24	37	686.9797
Jupiter	778833000		9	50	4332.62
Saturn	1426978000		10	14	10759.06
Uranus	2870991000		16	10	30707.79
Neptune	4497070000		18	26	60199.63
Pluto	5913510000	6	09	18	90777.61

* Retrograde (i.e., opposite the direction of the Earth's spin)

God is Eternal *(Al-Bāqī* الباقى). He was there before time began and He will be there when time ends. Most astrophysicists presume that before the beginning of time there was nothing. Suddenly, a "big bang" occurred and time began from a point of singularity. One second after this explosion, a colossal amount of heat (ten billion degrees centigrade) was still present. This energy started a chain of creational events. Elementary bodies, parti-

Science Miracles: No Stick or Snakes

cles, and electrons were formed first. Then atoms of hydrogen were produced which turned into smoke to form the nebulae from which stars, suns, and the planets were made. The great amount of heat must have originated from the primeval explosion. The Qur'an suggests that humankind was subsequently created from "sticky clay."

> Just ask their opinion: Are they more difficult to create, or the (other) beings We have created? Then have We created out of a sticky clay! (37:11)

فَاسْتَفْتِهِمْ أَهُمْ أَشَدُّ خَلْقًا أَمْ مَنْ خَلَقْنَا إِنَّا خَلَقْنَاهُمْ مِنْ طِينٍ لَازِبٍ (١١)

Surat al-Ṣāfāt

The Oort Cloud

Jan Oort, a Dutch astronomer, proposed in 1950 that the explosion which produced the Sun and the solar system ended in the formation of the Oort Cloud which extends 1173.33 times the distance between Pluto and the Sun — 5,913,510,000 km. The radiation from the Sun drove trillions of tons of material deep into cold space where it remains frozen until now. This material is locked into great clumps of icy dust that are the remains of the original building blocks of our solar system. David Levy, comet discoverer, describes a comet as a snowball about 10 km in diameter with a lot of dirt in it. This is what the Oort Cloud is made of and is essentially unchanged from the time the solar system was started. Larry Labuveski, replicated a comet, using dry ice, ammonia, some dirt, organic material and water. These are some of the most primitive compounds that were present during the formation of our solar system and when combined they make the dirty iceballs that we know as comets. The comets are the ghosts of our past, echoes of the earliest material in space. Normally, the Oort Cloud is the home of the comets (see Figure 1).

Extension of Earth and the Lithosphere

God described the creation of the Earth in different parts of the Qur'an. In *Surah Qāf,* He states that He is spreading it and He has thrown on top of it *rawasi* (lithosphere), plates which carry the continents and are in continuous motion. We now know that they are always extending at the oceanic ridges where two thirds of the world has been made over the last two hundred million years (50:6–7) (Abbas 1997).

Creation of Man

He created man (50:16) at first from a selected strain of mud (23:12), then his progeny from a selected strain of "despised" water (32:8). The word *sulalah* (selected strain) indicates a long process of propagation and selection. This deserves speculation and reflection as He created us in stages, *aṭwār* (71:14).

Possible Weight of Our Spirit (*Rūḥ*)

God states that whenever someone dies, the Earth will be reduced to a certain degree, we presume in weight. This suggests that when man dies his body mass will remain on Earth, but that an immeasurable quantity representing the energy responsible for life will disappear. Presumably this refers to the *rūḥ* or spirit. Such an immeasurable quantity is recognized and measured by The Omnipotent (*Al-Muqtadir* المقتدر) and recorded in a well-protected book.

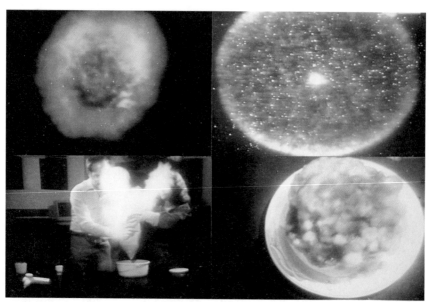

Figure 1: (TL) *Explosion to make the Sun and solar system.* (TR) *The Oort Cloud.* (BL) *Experimental formation of comet material.* (BR) *Artificial comet material, "dirty muddy ice."* (Discovery, September, 1997).

We already know how much the earth is reduced (when they die); and with Us is a record guarding (the full account). (50:4)

قَدْ عَلِمْنَا مَا تَنْقُصُ الْأَرْضُ مِنْهُمْ وَعِنْدَنَا كِتَابٌ حَفِيظٌ (٤)

Surah Qāf

Nearer Than the Venus Return to the Heart

Because He created us, He knows our thoughts. The word *wiswas* portrays how the Devil influences man. This may explain our feelings, in the subconscious mind (the id) which aim to satisfy our instincts that seek constant pleasure and self-preservation. God portrays how near He is to us by indicating that He is nearer to man than his own *vena cava* which returns the blood to his heart (50:16). It seems that there are built-in systems within us that observe us all the time.

Two Receivers Set on the Right and Left

One scientific allusion that may explain how God watches us from within is the presence of two receivers stationed on our right and left. They receive what we think, say, or do (50:17–18). One scientific explanation (Abbas 1997) suggests that such receivers are present in our brain. There are numerous brain centers with unknown functions, anyone of which could have been programmed by God to perform this function and we would be completely unaware of its existence. It has been shown in the same Qur'anic reference that God's scoring system favors man. He does not record evil thoughts, only if they are acted upon and then it is scored as one wrong deed only. On the other hand, even a good thought is scored as one good deed. All our good deeds are scored in duplicates and even in special circumstances they are scored many times their worth (6:160; 2:261). God is generous but also He is Watchful (*Al-Raqīb* الرقيب). Nothing said can be changed (50:29). God has no gain in punishing us (4:147).

Provision and Rehydration

God created plants using rainwater. Plants provide for animals and man (50:7,9–11). This is the food cycle by which He sustains life (*Al-Razzāq* الرزاق). Rainwater brings life to barren land, thus simulating how man is resurrected from dust. Human knowledge stops at suspended animation or lyophilization. It is a process in which live bacteria are suspended in a vacuum in ampules. To bring them to life, water is added. A drop of the suspension could produce billions of organisms if cultured on a suitable medium. The process is used in all laboratories to produce vaccines and to diagnosis diseases. It is also used in biological warfare where micro-organisms are stored in a lyophilized state, especially the resistant spore-forming types, and are spread by aerosol as a cloud or fog causing mass destruction.

When a body is buried, the bones and human remains stay together in the grave. Even the parts consumed by worms or bacteria that are present inside the body eventually die and remain at the site as matter. God describes how he brings life to barren land by rain and how he would resurrect man.

The Universe's Natural Phenomena

God, The Forewarner (*Al-Muqaddim* المقدم), repeatedly warns us in the Qur'an of the dangers of the universe's natural phenomena with which He punishs those who are destructive. He gives many examples of flood, fires, earthquakes, bolts of lightening, acid rain, hurricanes (50:12,13,14,36). These disasters may be Earth-initiated or triggered by meteorites, asteroids or comets from outer space (the subject of a later chapter). Since He is The Equitable (*Al-Muqsit* المقسط), He will not punish until He has given sufficient warning through a prophet or the Qur'an (50:2,45).

Life and Death and Supra-Sonic Waves

If we accept that God created the heavens, the Earth, and what is between, then it should not be difficult to accept that He can repeat the creation again if He so wishes (50:15). One certainty exists in our life, and that is death, which is brought about by The True *(Al Haqq* الحق), The Collector of our Life *(Al-Qābiḍ* القابض) (50:5,19,42). He is The Supreme *(Al-Qahār* القهار) who will use sonic waves to visit death (50:20) on us on the Day of Doom and again to resurrect us (50:42). According to our scientific knowledge, sonic waves can only be used for destruction. We use it to kill bacteria to prepare vaccines; we are not aware of a constructive use of supra-sonic waves — certainly not to initiate life. God, The Avenger *(Al-Muntaqim* المنتقم) will avenge those who dispute His existence when they can see evidence of His creation all around them (50:20,24).

Science Beyond Time

God provides us with some knowledge relating to the life-to-come, which may explain some of our psychological behavior. It may also shed some light on our spirit or self. He says that we have companions with us all the time, some of whom are capable of leading us astray. At the time when all souls are held accountable, they will deny their role in our temptation. Their arguments and ours, all of which were recorded during our lives, will be produced as evidence. Once recorded they cannot be changed (50:29).

After our resurrection, each self is led by a driver and accompanied by a witness. Angels are presumed to do this duty; however, this statement may have a far-reaching scientific explanation (50:21).

On the Day of Resurrection, we shall hear "the cry of Truth," from a "near place." The Earth will split into successive cracks, perhaps through earthquakes, and humans will erupt from them swiftly. An easy gathering for The Holy One (*Al-Quddus* القدوس) (50:41–42,44). All of these allusions are mysterious, and they indicate a powerful creator.

CHAPTER FOUR

PATIENCE IS A VIRTUE

In this chapter we will study how God teaches man the virtues of patience. He is absolute in power, yet He exercises great patience toward all His creatures, including the Devil. Nothing can be accomplished without patience. God has been patient and will remain patient even toward those who have claimed lies against Him for several millennia. He will bide His time until the promised Day of Judgment. Unfortunately His name, The Patient *(Al-Ṣabūr* الصبور), does not occur in the Qur'an but is known by implication. Perhaps it is for this reason He begins a *surah* with the letter *ṣād* which is the initial of the name *Ṣabūr* (Patient). In many verses God asks people, including the prophets, to be patient in the knowledge that they will receive their reward in heaven. He orders the Prophet Muhammad to be patient, indicating that his patience is from God (16:127). This infers His name The Patient.

And be patient, for your patience is but from Allah; nor grieve over them: And distress not yourself because of their plots. (16:127)	وَاصْبِرْ وَمَا صَبْرُكَ إِلاَّ بِاللَّهِ وَلاَ تَحْزَنْ عَلَيْهِمْ وَلاَ تَكُ فِــي ضَيْقٍ مِمَّا يَمْكُرُونَ (١٢٧) Surat al-Nahl

The Symbolic Letter Ṣād (ص)

The letter *ṣād* appears at the beginning of the first verse of *surah* 38 (the thirty-eighth *surah* cronologically), which takes its name from it. Surah Ṣād contains 88 verses. The letter occurs infrequently in the *surah*, only 29 times out of 3,061 letters (0.94%). It stands out, however, when God orders the Prophet Muhammad to "be patient" (*iṣbir* إصبر) when people doubt his message and his authenticity (38:4–8,14).

God has only two names which begin with the letter *ṣād*: The Patient and The Absolute (*Al-Ṣamad* الصمد) (112:2). Explanation of the letter *ṣād* can easily be covered in the light of those two names. God begins the *surah* stating that the Qur'an is not only for remembrance and reading (38:1) but also for understanding, meditation, and to work out the meaning of its verses (38:29).

God Provides Absolute Protection to His Message on Earth

God created the heavens and the earth and what is between them (38:27). He sent His prophets, from Noah to Muhammad (12:13,14,17), with one instruction: to be safe from man's hand and tongue, to believe in God, to become His Khalifa on Earth (27:62), and to develop it by reflection on God's physical sciences as revealed in the Qur'an. God is not prepared to allow anyone to fight against His cause or to defeat His subjects (38:3,27–28). He has defeated soldiers of destruction in the past (38:11–14) and He has prepared for them severe punishment in the life to come (38:55–64). Those who succumb to the Devil's temptation receive a similar fate (38:84–85). He possesses the means of mass destruction, some of which are the universe's natural phenomena. In the past, He used them to destroy those who corrupted the Earth. In Moses's time he used drowning; in Noah's, flooding; in Hūd's, the hurricane; in Thamūd's, earthquakes; in Lot's, brimstone; and in the time of the Companions of the Wood, volcanic eruption. (Further details will come later.) God exercised a great deal of patience and mercy during his punishment (38:9).

The Lessons of Patience

In *Surah Ṣād*, God shows the Prophet Muhammad how He, the Creator, with all His might and power, possessing all methods of mass destruction in the universe, is still forgiving (38:66). He set Himself as an example giving the Prophet Muhammad several situations in which He exercised patience with people, prophets, and even with the devil.

God's Patience with Ordinary People

God was very patient with the peoples of Noah, Moses, and the other prophets. He did not punish them until He showed them His signs, sent messengers to teach, preach, and warn before He finally punished them (17:15).

God's Patience with His Prophets

God bestowed many powers on David. The mountains and the birds used to pray with him. He also gave him the wisdom of making sound decisions (38:18,19,20). In one incident, however, he abused his power of judgment. The Prophet David had 99 wives, yet he took for himself another man's only wife (narrated by al-Tabari, see M. Asad's Qur'an translation). He claimed her for himself. Although his palace was guarded by thirty thousand strong soldiers, two angels in the shape of men climbed the wall and startled him in his quarters. They claimed they were brothers. They summarized their complaint by indicating that the first had 99 ewes, and the other had one. The first was claiming the single ewe to add to his 99 and asked David to adjudicate between them. David judged the brother with the 99 ewes to be guilty of greed. The angels disappeared and the Prophet David realized that he had sinned by taking his hundredth wife from another man. He dropped to his knees, asking God's forgiveness (38:21-24). God forgave him (38:25) and said to Him: "We made you our successor (*khalifa*) on earth. Judge people in truth and justice and do not follow the lusts (of your heart)." Those who are misled from the way of God are punished severely (38:26). God was patient.

The Prophet Solomon, son of the Prophet David, forgot God in his love for horses (38:31–33). He also fell in love with a woman who worshipped idols. He possessed a ring, a symbol of his kingdom and authority over humans and jinn, which he used to take off during his sleep and give to this woman. She gave the ring to a jinn who assumed Solomon's appearance and sat on his throne (38:34) (see Ibn Kathir). When the Prophet Solomon realized what she had done, he reclaimed his ring, his kingdom, and his authority over man and jinn. God was patient with Solomon and forgave him. He gave him the power to rule the wind and the jinn (38:36–38).

God's Patience with the Devil

God created man from mud and asked the angels to kneel to him. They all knelt except the Devil. The Devil was thrown out of heaven, yet God did not punish him. The Devil asked God to respite him until the Day of Resurrection. God said: "Thou art among the ones who are granted respite" (38:71–81). God is patient even with the Devil.

The Symbol of Patience

God set another example of patience for the Prophet Muhammad, the Prophet Job. After all his sufferings, God rewarded him for his patience by returning him to full health and by resurrecting his dead children (38:41–44).

God asked the Prophet Muhammad to be patient with his people. He came to warn them and give the great news of the Qur'an (38:65–68,86).

CHAPTER FIVE

THE FIVE 'W'S: WHERE, WHY, WHO, WHEN AND WHAT

Since man's existence on Earth, God has promised to send him guidance, which he should wait and look for (20:123). *Surah Ṭāhā* explains one of His main systems of guidance to man: a prophet, chosen from among his own people so that he may be acceptable (9:128). The system answers the five questions beginning with "W" which we use for investigation and finding solutions for population problems.

The Symbolic Letters *Ṭā Hā* (طه)

The letters *ṭā hā* constitute verse 1 of the *surah* that is named after them and contains 135 verses.

Traditional Explanation
The common belief among Muslims is that *Ṭaha* is a name that God gave to the Prophet Muhammad. In verse 1, it sounds as if He is addressing him by the name of *Ṭaha* and in verse 2 He says that the Qur'an has not descended for his suffering (20:2,3). Many Muslims call themselves *Ṭaha*. A publishing house in London, specializing in Islamic publications, does business under the name *TAHA*. Most of *surah* 20, however, is concerned with the Prophet Moses (9–99) and Adam (115–133).

Computer Search
A computer search of the Qur'an for words beginning with the two letters *ṭā hā* reveals only three words. The first occurs at the beginning of *surah* 20 as we have mentioned. The second word is *ṭahira* طهرا (2:125). God uses this word to ask the Prophets Abraham and Ishmail to purify His

house, the Ka'ba. The third word is *tahūra* طهورا, which means "pure" and is used to describe rainwater (25:48). The only interpretation that can be made on the basis of words beginning with *ṭā hā* is that the Prophet was described, perhaps, as being pure. But this explanation does not sound satisfactory.

Interpretation in Relation to Other *Surah*s and the Names of God

The symbolic letter *ṭā* (ط) appears in the Qur'an in three more *surah*s 26, 27, and 28 (47,48, and 49 chronologically). The four *surah*s that contain *ta* also give full details of God's guidance to man through Moses. The *surah* under study (20), narrates the story in 90 verses (20:9–98). *Surah* 28 completes the story in another 41 verses (28:3–43). In both *surah*s God says that He called Moses from the right side of Mount *al-Ṭūr*. The word *Ṭūr*, which refers to the sacred mountain, begins with the letter *ṭa*. Perhaps God started the two *surah*s with the letter *ṭā* to indicate the first occasion when He spoke to a human being, an important step to man's guidance and the sacredness of the place.

God, being The Benign (*Al-Laṭīf*) and the Merciful, guided man instead of leaving him in darkness. He kept His promise by sending prophets. Perhaps God put the *ṭā* at the beginning of *surah*s 20 and 28 to indicate the relationship between the story of Moses in each. The story shows *where* and *why* He sends a prophet and *when* and by *what* technique they carry on their message. The following verse shows that the ultimate objective of Moses as a prophet is to deliver the mercy and contained guidance in His Book, the Torah (28:43).

We did reveal to Moses the Book after We had destroyed the earlier generations, to give insight to men, and guidance and mercy, that they might receive admonition. (28:43)	وَلَقَدْ آتَيْنَا مُوسَى الْكِتَابَ مِنْ بَعْدِ مَا أَهْلَكْنَا الْقُرُونَ الْأُولَى بَصَائِرَ لِلنَّاسِ وَهُدًى وَرَحْمَةً لَعَلَّهُمْ يَتَذَكَّرُونَ(٤٣) Surat al-Qaṣaṣ

Explanation of *Ṭā Hā* (طه) in the Light of God's Names

- The *ṭā* (ط) is part of three names of God: The Benign (*Al-Laṭīf* اللطيف), The Innermost (*Al-Bāṭin* الباطن), and The Equitable (*Al-Muqsiṭ* المقسط).
- The letter *hā* (ه) is the beginning of His name, The Guide (*Al-Hādī* الهادى).

Since Adam inhabited the Earth, God, being The Benign and The Merciful, has tried to guide man (16:64; 28:43; and 20:2–3). Being The Innermost, He knows the secrets of the universe, including what is inside the earth (20:4–6). He is also aware of man's inner feelings and thoughts (20:7). He gave us His beautiful names to help us pray to Him and perhaps help us to understand the secrets of the Qur'an (20:8; 7:180; 17:110; 59:24). Being equitable He does not punish man without proper education and teaching. He ensured this by sending messengers, in the form of prophets, before He inflicts any punishment (28:59).

The Guide (*Al-Hādī* الهادى)

God guides man by means of the most sophisticated scientific approach. The letter *ṭa* directs us to use both *surah*s 20 and 28 to understand how God guides man by means of prophets. The system follows our modern methods of research of epidemiological problems in which one has to answer the five questions beginning with 'W'. **Where? Why? Who? When?** and **What?** Where and Why we need the guidance most? Who is going to guide? When He is going to do it and by What technique? The objective of the process is guidance (20:135).

Prophet Moses — An Example of Guidance

Where? and Why?

God, The All-Seeing (*Al Baṣīr* البصير), saw that Egypt was a place where a king (Pharaoh) claimed he was god. He maltreated the people in his land; he killed their sons and spared their women (28:3,4).

Who?

Choice of Prophet
God chose the Prophet Moses as a child (28:7–13). He educated him in the highest place of his time, at Pharaoh's palace (28:9; 26:18). He watched over him and forgave his mistakes (28:15–28; 20:40).

Giving Him Signs
God called the Prophet Moses to show him His signs (20:11; 28:46). He gave him his miracles, to prove that he is a genuine messenger from God (20:17–23).

Giving Him Support
God removed the Prophet Moses's stutter (20:5–28) and supported him with his brother by making him a prophet.

When God Decided It Was Time
Moses was given his task, to spread God's guidance to Pharaoh, his soldiers, his administration and the people of Israel (20:24).

The Technique the Prophet Moses Used to Spread His Message

At first, the Prophet Moses approached the people, preached and guided them using the knowledge given to him by God (20:42–55). When they were in doubt he showed them God's signs or miracles (20:56). His heavenly miracles defeated human imitations (20:57–69). Some of the people believed in him (20:70–76) and some disbelieved (20:71). Moses showed them further signs from God, nine in total (17:101 & 27:12). Still Pharaoh and his followers disbelieved in him and his message.

God inspired Moses to take the believers (the Children of Israel) out of Egypt. Using another sign from God he split the waters, they crossed safely while Pharoah's followers were drowned (20:77–78).

After all this guidance, the people of Israel still denied God (20:79–97) and worshipped a statue (20:88–89). Moses asked God to forgive them, then he chose seventy men from the people of Israel to meet God. They were seized by an earthquake. The Prophet Moses asked God not to punish all of them for what the foolish ones amongst them had done (7:155). He said they have repented (7:156). In Arabic the word "repented" is "hudna" هدنا. This is probably the origin of the word *yahūd* (يهود) in Arabic (Jew).

Guidance Since Adam

Since Adam inhabited the earth (20:115–121) God has favored, forgiven, and guided him. He said that people would become enemies to one another and that they should expect guidance from Him: Those who follow His guidance will never go astray or be unhappy (20:1–3,135). We showed how He guides people through messengers sent as prophets by answering the five questions beginning with "W" in *surah*s 20 and 28. In Surah 20, Allah informs us that He sent prophets who followed the same course, who were part of a great heavenly program directed from The Guide *(Al-Hādī* الهادى).

The Objective of Guidance

The continuous message of guidance aims at establishing man as God's successor on Earth. He must attain a high profile of morality and ability to qualify for such a high position. His performance, if it is within the guidelines of God, will enable him to reach a position with God through which he will obtain light that surrounds him on Earth and in the life to come.

CHAPTER SIX

LIFE, DEATH AND PEACE

High Mystical Value

Muslims consider *surah* 36 (*Surah Yasin*) to be of special spiritual significance. Many Muslims carry a copy of it together with other short *surahs* as a talisman to protect them. It has been said that if one reads it at night before sleeping he/she will not die until the morning. Many people read it each night before they sleep.

The Symbolic Letters *Yā Sīn* (يس)

These two letters constitute the first verse of *surah* 36 (41 cronologically), and from them the *surah* takes its name. It contains 83 verses.

Traditional Explanation of the Letters
When the letters *yā* (ي) and *sīn* (س) are combined they are read *yasin*. It is believed that this is a name given to the Prophet Muhammad by God. Muslims often name their children Yasin.

Computer Search
In a computer search, 115 words were found to begin with the letters *yā* and *sīn* but none of these words seem to have any bearing on the *surah* under study. Two words, however, were noted which could be of some significance. The word "yuslim" (يسلم; he submits [the present tense verb of the noun "peace," which is one of the names of God]) (31:22), and the word *yasṭurūn* (يسطرون; writing in lines) (68:1).

Explanation in the Light of the Names of God

Explaining the letters *yā* and *sīn* in the light of the names of God, yield an interpretation other than the name of the Prophet Muhammad. The letter yā occurs in 42 names of God and the letter sīn occurs in 6 names. All the verses of this *surah* can easily be explained in relation to any of these names; however, the names, chosen for interpretation, begin either with the yā or sīn. The main themes of this *surah* can be related to three names.

Power of Giving Life *(Yuhyi* يحيى*)*
and Causing Death *(Yumīt* يميت*)*

No matter how many prophets or heavenly signs come to people, they always have doubts about resurrection. They can watch the dead land burst into life: after a rain barren land will burst to life with plants, fruits, and seeds, feeding man himself and his animals (36:33–36). They can see the power of God in His signs: the alternation of the day and the night, the swimming of the sun and moon in their specified orbits, and the safe return of people from the sea who set out without the aid of navigation instrumentals to guide them (36:37–44). The creator who has all these powers can give life and cause death.

God states that He resurrects the dead (36:12); He gives us examples from the past. Two disciples went to preach in Antioch and a third went to support them. The latter was stoned to death. God resurrected him and placed him in heaven (36:26–27). God killed the people who stoned him with a supersonic wave (36:29). All the people who die will be resurrected (36:31–32).

God will use supersonic waves to bring death to people on the Day of Doom (36:48–50). He will also use supersonic waves to resurrect them (36:51,53). This scientific allusion has been discussed previously.

Peace (*Salām* سلام)

Peace is a name of God. In his heart, everybody seeks to be at peace with himself and with others. Muslims greet each other by saying, "Salam 'alaykum" (peace be with you). Islam originates from *Al-Salām*, a name of God. The definition of a Muslim is: a person from whom others are safe from his/her hand and tongue. Peace seems to be the objective of all nations, yet we are unable to live together in peace (7:24).

God gives *salām* to those who have been chosen to enter paradise (36:58). God addresses those who denied Him and His signs as criminals (36:59–67).

Peace does not appear to exist on Earth; however, man's goal should be to live in peace and to die in peace. During his life he should make his peace with God and believe in God's ability to resurrect him so that he may find peace in the life to come.

CHAPTER SEVEN

IMAGING

Our understanding of "imaging" has been primitive until recently. Imaging was recognized as a science with the introduction of fibre optics, magnetic resonance imaging, and radio imaging. Images from the Moon, planets and other objects in space are continuously sent by satellites, thus daily increasing our store of information.

This progress falls short of the depth of knowledge that goes into "imaging" man during his creation in the womb. More knowledge and technology is required to image animals, plants, and other objects on Earth and in the universe.

The Symbolic Letters *Alif Lām Mīm Ṣād* (المص)

These four letters constitute verse 1 of *surah* 7 (The Heights, *Al-A'rāf*), which contains 206 verses.

Computer Search

In the Qur'an, the four letters appear at the beginning of nine different words which appear in 34 verses (see Table 10).

Table 10: Words in the Qur'an That Start with Alif Lām Mīm Ṣād (المص).

Word	Word in English	Surah:Verse	Word	Word in English	Surah:Verse
المص	*Alif Lām Mīm Ṣād*	7:1	المصلح	The Gooddoer	2:220
المصباح	The Lamp	24:35	المصلحين	The Righteous	7:170; 28:19
المصدقين	The Charitable Men	37:52, 57:18	المصلين	Those Who Pray	74:43
المصطفين	Among the Elect	38:47	المصور	*The Imager*	59:24

Word	Word in English	Surah:Verse
المصور	Destination	2:126; 5:18 22:72; 35:18; 57:15; 64:10; 2:285; 8:16, 24:42; 40:3; 58:8; 66:9; 3:28; 9:73; 24:57; 42:15; 60:4; 67:6; 3:152; 22:48; 31:14; 50:43; 64:3
NB. Please note the highlighted name of God, The Imager.		

The most outstanding, and by far the most important word among them, is God's name, the Imager (*Al-Musawwir* المصور). In *surah* 7, God asks people to pray to Him using His beautiful names (7:180), which include the Imager (59:24). The scientific implications of the name have been appreciated only recently. Imaging animals, people, and all the objects in the universe must be very complex. Perhaps God put the letters *alif lām mīm ṣād* (المص), which are the initials of His name, The Imager, at the beginning of *surah* 7 to mark the verses that show His power of imaging in creation. Imaging is a complex procedure, any fault could cause malfunction, resulting in strange looking monsters or distorted creations. *Surah* 7 gives a few examples to show the power of imaging during the creation of man.

Creation and Imaging of Man

God created man and "imaged" him (7:11). The technique of imaging man in the womb until full term and beyond is probably going to remain one of the many secrets that God will keep to Himself. We have reached, however, a degree of knowledge that may help us appreciate the complexity of this process and its difficulty. In the following paragraphs some of the recently discovered scientific procedures and findings are given.

It is We who created you and gave you shape; then We bade the angels bow down to Adam, and they bowed down; not so Iblis; he refused to be of those who bow down. (7:11)

ولقد خلقناكم ثم صورناكم ثم قلنا للملائكة اسجدوا لآدم فسجدوا إلا إبليس لم يكن من الساجدين (١١)

Surat al-'Arāf

Magnetic Resonance Imaging

A detailed explanation of the scientific implication of this process has been attempted elsewhere (Abbas 1997). The process of "imaging" in man is still in its infancy. We know that man's body, like the body of an animal or plant, is composed mainly of water. Performing Magnetic Resonance Imaging (MRI) on a human, depends on imaging the resonance of the protons present in hydrogen, which is part of the water molecule. As a result, the proton will resonate to a radio frequency pulse in a magnetic field. Body organs contain different amounts of water and will subsequently have different amounts of hydrogen; therefore, different amounts of hydrogen protons are present in different organs. The resultant resonance from each organ will give an unique image. Images of normal organs are then compared with images from abnormal ones (see Figures 2 & 3).

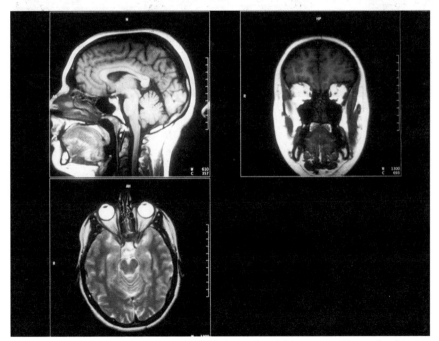

Figure 2: Magnetic Resonance Image. (TL) Sagittal mid-line section T1 weighted image showing a normal anatomy. (TR) Coronal section through the anterial part of the head showing the frontal lobes of the brain, the back of the eye socket (showing the nerves and muscles) nose, mouth, and tongue. (B) Axio T2 weighted scan through the mid-cranium clearly showing part of the brain, the brain stem, and the orbits.
(Courtesy of Dr Paul Spencer, BSc, MRCP, FRCR, Consultant Radiologist, Rotherham District General Hospital Trust)

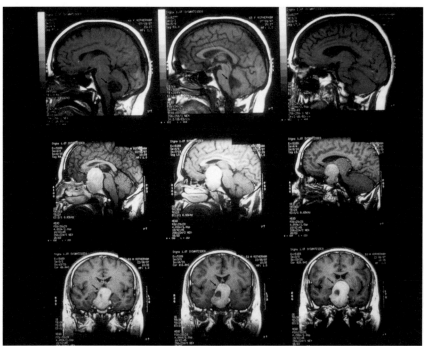

Figure 3. Magnetic Resonnance Image. **(TR-TL)** Sagittal T1 images demonstrating a large cystic mass in the cerebellum. MRI is particularly useful in showing tumors and abnormalities in the deep part of the brain. Sagittal **(MR-ML)** and coronal **(BR-BL)** T1 weighted contrast enhanced sections through the pituitary region, showing a large pituitary tumor. MR imaging clearly outlines the full extent of such tumors, which guides surgery.
(Courtesy of Dr Paul Spencer, BSc, MRCP, FRCR, Consultant Radiologist, Rotherham District General Hospital Trust).

Mirror Image

God images us in three veils of darkness, which presumably represent the foetal membranes (39:6). He makes one-half of our body a mirror image of the other. The right side of our face is an almost identical mirror image of the left side at the middle line. Similarly the right side of our body is a mirror image of our left side. Subsequently, our hands and feet are mirror images of each other; otherwise, we would have two right feet and two right hands or vice versa.

Genetic Map

From the very beginning of man's creation, soon after the combination of the sperm and the ovum to form the zygote, his cell has a genetic map that will repeat itself in all the subsequent cells that are produced to replace dead cells or during growth. This genetic map is unique to each individual and is formulated from genes made from DNA (Deoxyribonucleic Acid), which is made up of a group of amino acids in different arrangements that make them unique for each person. DNA is used in the identification of doubtful paternity and for proving criminal charges.

The Human Genome Project

There is a world-wide effort to generate an encyclopedia of our genes. The project, it is hoped, will be completed between 2002 and 2005. It will give details of physical maps of human chromosomes and the overlapping DNA segments along these chromosomes. The latter can be used to locate novel segments of DNA and can cut the time taken to locate genes from years to months. The project will provide vital clues for the prevention, early diagnosis, and cure of many diseases. The problem is: How can we make the best use of such an enormous and detailed database on human genetic material?

Knowing the DNA sequence of a gene, where on the chromosomes it can be found, and in what cells it is active, is only a small part of understanding how its product works in the body. Genes do not function in isolation. The entire set is integrated. There are layers of genes interacting with and controlling the action of other genes. Different groups of genes at different times and in different places are switched on and off to manufacture the proteins that operate cells. We have not yet cracked the important code of how the amino acid sequences lead to protein function.

We know there are special genes that predispose to specific diseases, e.g., glaucoma (increased pressure within the eye). Twenty genes are associated with hereditary forms of cancer.

The subject is extremely complex. For example, researchers have explored the possibility that leukemia (white blood cell malignancy) may be caused by some fault in the control of white blood cell formation. During the research a regulator called "colony stimulating factor" (CSF) was discovered, and it is now used in the treatment of this disease. A later discovery showed that each of the eight types of white blood cells is regulated by seven or eight growth factors of which the CSF is only one. We do not know where these growth factors are made in the body or how they exert their control over the formation of white blood cells. They give instructions to the cells like: begin to divide, make this particular product, mature, and move elsewhere. We also do not know the steps in the signal pathway. We do not know how cells speak to one another or what they are trying to tell us.

Programmed cell death is part of the normal regulatory process. When cell division becomes uncontrollable, or if cells simply fail to die, we develop cancer.

With new developments in our knowledge and science we are hoping to teach disfunctional cells healthier behavior. Some biotechnology companies are already using genome typing "chips" to look at inheritance patterns. We are hoping to locate genes that cause diseases and "knock them out" of action.

It is hoped that the Human Genome Project will produce something similar to the periodic table in chemistry in its usefulness. Though the project is all-encompassing in scope, requiring immense work, it will never yield a complete picture of the human image. Because of its complexity, using it will require computer experts, as well as scientists in genetics, molecular biology, biotechnology, medicine, and many other fields of science.

Human Time Travel Transfer

In science fiction films, we see how people are transferred by radiation from Earth to space and from the past into the future and vice versa. They call this process "beaming," in which matter becomes vaporized and "beamed" to wherever the producer wishes. They use an imaginary machine which they call Matter Energy Converter. The producers of these films, our young scientists, and science fiction film fanatics should be extremely intrigued with what God did to us before our creation. Somehow God "beamed" all human beings, including those who were not yet born, to His presence. He asked us to testify in front of Him and ourselves that He is the Creator, which we did. He wanted to make sure that we do not claim ignorance (7:172). Now He is sending us through creation. Some have already been born and died and many are still to be born, and the process will go on until the Day of Doom. Many scientific questions can be put forward. Did we appear as energy? Did we appear in body? Were we there before we existed? Millions of questions may be asked. Only the Imager has the answer.

> When thy Lord drew forth from the Children of Adam — from their loins — their descendants, and made them testify concerning themselves (saying): "Am I not your Lord (Who cherishes and sustains you)?" they said: "Yea! We testify!" (This), lest ye should say on the Day of Judgment: "Of this we were never mindful." (7:172)
>
> وَإِذْ أَخَذَ رَبُّكَ مِنْ بَنِي عَادَمَ مِنْ ظُهُورِهِمْ ذُرِّيَّتَهُمْ وَأَشْهَدَهُمْ عَلَى أَنْفُسِهِمْ أَلَسْتُ بِرَبِّكُمْ قَالُوا بَلَى شَهِدْنَا أَنْ تَقُولُوا يَوْمَ الْقِيَامَةِ إِنَّا كُنَّا عَنْ هَذَا غَافِلِينَ (١٧٢)
>
> Surat al-'Arāf

Human Extension

Another fascinating process of imaging is God's creating Eve from Adam (7:189). We know yeast cells multiply by budding. Plants can multiply by a similar process. How did a woman come out of a man? Everything is possible for The Imager.

> It is He who created you from a single person, and made his mate of like nature, in order that he might dwell with her (in love). (7:189)
>
> هُوَ الَّذِي خَلَقَكُمْ مِنْ نَفْسٍ وَاحِدَةٍ وَجَعَلَ مِنْهَا زَوْجَهَا لِيَسْكُنَ إِلَيْهَا (١٨٩)
>
> Surat al-'Arāf

In the next chapter, we shall see examples of creation of man by a process reminiscent of parthenogenesis and other techniques. He is the master of human genetics. God confirms His power of imaging in creation when He states that He made us out of water with very little solids, yet we might be males, females, tall, short, dark, or fair (25:54). He also made animals and plants out of water (24:45; 21:30). This is why He is the Imager.

CHAPTER EIGHT

PARTHENOGENESIS AND REGENERATION OF ORGANS

In the previous chapters, we have seen how God creates life from death not only in barren soil but also in human beings. We have seen how He, as an imager, creates man and images him in two symmetrical halves, giving him a genetic map identical for every cell in the same person and different in each person. We have seen how it will take the scientist almost twenty years to draw (an incomplete) genetic picture of man: A project that we call the Human Genome Project. We are trying to learn how to "knock out" genes that predispose us to diseases like glaucoma and cancer.

In this chapter, God will show us how He creates life and regenerates nonfunctioning ovaries, uteri, and probably testicles to bring back fertility. God will also show us that He can generate reproduction without sexual intercourse: a process that we call parthenogenesis. God makes it happen in nature all the time, in birds, reptiles, insects, and plants. A similar, modified process happens at all ages in human females and produces what we call teratomas.

The Symbolic Letters *Kāf Hā Yā 'Ayn Ṣād* (كهيعص)

These five letters formulate the first verse of *surah* 19, Maryam, which contains 98 verses.

Computer Search

No word in the Qur'an begins with these five letters. Each of these letters might be present in the names of God that the Prophet Zachariah used during his secret prayer to God (19:2,3). A prayer containing possible names may be as follows:

O God, who is generous, bigger than all, the Guide who gives life and causes death, who is Great, Mighty, and Knowledgeable, All-Knowing, High, Just, Forgiving, Patient and Absolute, bestow on me a child although I am old and my bones are brittle, and my wife is very old and barren (19:4–6).

سبحانك اللهم الكريم الكبير الهادى الذى يحيى ويميت العظيم العزيز العليم العادل العفو العلى الصمد الصبور لقد كبر سنى و وهن عظمى وامرأتى عجوز عاقر اللهم ارزقنى طفلا يرثنى.

The prayer could have been simple with only five of God's names, each containing one of the symbolic letters. On the other hand it could have contained all the possible names and Prophet Zachariah's request, for example:

O God, who is the Generous, be generous to me. O You who are greater than all, raise me among my people by giving me an heir. O You who are the Guide, I fear my faith will disappear after my death, so give me a child to carry on my message. O You who are the Giver of Life and Causer of Death, put life in my old body and help my brittle bones. Put life in my old, barren wife and put life in her withered organs. O You who are the Great, shade me with Your greatness. O You who are the Mighty, help me to overcome my weakness. O You who are the All-Knowing, help us with Your knowledge. O You who are The Just, we have no successor, we beg You to adjust. O You who are the Forgiver, please forgive us. O You who are the Absolute, give us a child to continue Your guidance and Your message. O You who are The Patient, reward my patience and my humility among my people and give me a successor.

يا كريم اكرمنى، يا كبير ارفع من شأنى بين قومى واعطنى وريث، يا هادى أخشى أن تندثر رسالتى من بعدى فارزقنى وريثاً يكمل رسالتى، يا من يحيى ويميت ابعث الحياة فى جسدى وعظامى الواهنه، وضع الحياة فى أعضاء زوجتى العاقر، يا عظيم اشملنى بعظمتك، يا عزيز ساعدنى فى ضعفى يا عليم عاونى بعلمك، يا عادل ليس لدى من يتبعنى فانصفنى، يا غفور اغفر لى، يا صمد ارزقنى بطفل يحمل رسالتى من بعدى، يا صبور عوض صبرى بمن يورثنى.

Explanation of the *Surah* in the Light of the Beautiful Names of God

One of the main themes of this *surah* is that God not only causes death (*Yumīt* يميت) but He also a gives life (*Yuhyī* يحيى). His knowledge is

Science Miracles: No Stick or Snakes

absolute (*Al-Alim* العليم); He demonstates how He can produce life by many techniques. He initiates it from nothing (19:9,66,67). Other methods are given below.

Regeneration of Senile Organs

God has the knowledge and power to regenerate a nonfunctioning ovary to produce ova and to regenerate nonfunctioning testicles to produce sperm. In the case of the Prophet Zachariah, his wife was very old and barren. Every girl is born with about 300,000 ova. After puberty one of these ova is shed off from the ovary during each menstrual cycle. In the case of twins or multiple pregnancies, two or more ova were shed from one or both ovaries. At menopause, such ova would have been used either by eruption out of the ovary or by transformation to other degenerative stages. Similarly, if the Prophet Zachariah was old and his bones brittle, it is quite possible that his testicles were no longer functioning. His semen would be either sterile or, his sperm count could have been so low or of low quality, to be unable to swim to reach the ovum, inside the abdominal cavity. The journey the sperm takes to reach the ovum is, in comparison to its size, similar to a man trying to reach the moon. The sperm have to overcome obstacles which hinder its journey, e.g., mucus etc.

God, with all His knowledge and capabilities, can regenerate the ovaries and the testicles to allow the Prophet Zachariah and his barren wife to have a child (19:7).

Recently, doctors are attempting to produce fertility in menopausal women. They give women who have reached menopause, hormonal therapy to revive ovulation. They then remove the ova, fertilize them with the husband's sperm in a test tube, then replace a fertilized ovum into the uterus. They continue to support the woman with hormones to maintain a continued pregnancy until the child reaches full term. This procedure is not accepted in England. If a doctor can perform this procedure, it must be elementary for the Creator. God gave life to all human beings. Had He wished, He could have made our treatment fail. Perhaps God allowed this practice to succeed to show us that what He did with the Prophet Zachariah was very easy for Him. He had already done it before, He gave the Prophet Abraham his sons Ishmail and Isaac (19:49). He is always The Generous (*Al-Karīm* الكريم).

Reproduction by a Process Similar to Parthenogenesis

God's creation of the Prophet Jesus in his mother is reminiscent of the process recognized as parthenogenesis (reproduction without sexual union) (19:19–23). Parthenogenesis occurs in some birds, reptiles, insects and plants. In man, women frequently present themselves to their doctor with a cyst on the ovary. This can be small but, if left, it can reach the size of a child. It is called Dermoid Cyst or Teratoma. Generally, cells multiply to produce identical cells. It was thought that for the ova to differentiate into the three types of cells, ectoderm (the future skin and brain) mesoderm (the future muscle and bone) and endoderm (the future intestine) it needed to combine with the male sperm. To form the zygote, this combination needs to trigger division and differentiation into the three embryonic layers; however, this is not needed for the formation of teratomas. Every week in hospitals, we remove teratomas from ovaries of women. Such teratomas originate from the ovum. These cysts may contain any type of tissue: teeth, hair, bones, lung, or intestine (see Figure 4). It has been accepted that if such ova migrate into the uterus by mistake, they would not continue to develop into a child; however, a report from Scotland shows that the genes of the white cells from a child are almost all identical to its mother and not its father (Bonthron, D.T. *et al.* 1995). It should not be difficult for God to create a child from an ovum of its mother without sexual union. This would be a simple procedure compared to creating the human genome.

Similarity of Regeneration and Parthenogenesis

Prophet Zachariah's child became a prophet by the name of Yahya (يحيى). God gave him this unique name. Its origin is from His own name, Yahya (يحيى), which means "giver of life." The two names are identical in writing but, there is a slight difference in their pronounciation. He created Yahya by regeneration of nonfunctioning organs. In the case of the Prophet Jesus, He created life from a living ovary without sexual union.

Jesus and Yahya will have the same life cycle: "They were born, they died and they will be resurrected" (19:12,15,33,34). It is not correct to call one "Son of God" and not the other. Both were created by His Power and by more or less related techniques. Both were prophets.

Science Miracles: No Stick or Snakes

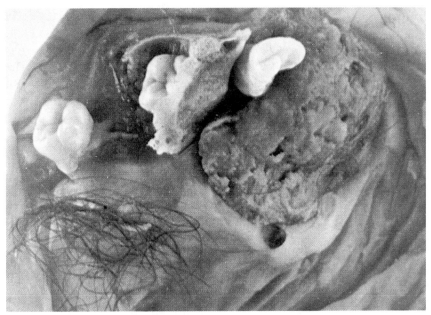

Figure 4: *Tertoma or dermoid cyst. Please note the presence of teeth, hair, and other tissues.*

(To his son came the comand) "O Yahya! take hold of the Book with might." And We gave him wisdom even as a youth. (19:12)

يَايَحْيَى خُذِ الْكِتَابَ بِقُوَّةٍ وَءَاتَيْنَاهُ الْحُكْمَ صَبِيًّا(١٢)

Surah Maryam

So peace on him the day he was born, the day that he dies, and the day that he will be raised up to life (again)! (19:15)

وَسَلَامٌ عَلَيْهِ يَوْمَ وُلِدَ وَيَوْمَ يَمُوتُ وَيَوْمَ يُبْعَثُ حَيًّا (١٥)

Surah Maryam

"So peace on me the day I was born, the day that I die, and the day that I shall be raised up to life (again)!" Such (was) Jesus the son of Mary: (It is) a statement of Truth, about which they (vainly) dispute. (19:33–34)

وَالسَّلَامُ عَلَيَّ يَوْمَ وُلِدْتُ وَيَوْمَ أَمُوتُ وَيَوْمَ أُبْعَثُ حَيًّا(٣٣)ذَلِكَ عِيسَى ابْنُ مَرْيَمَ قَوْلَ الْحَقِّ الَّذِي فِيهِ يَمْتَرُونَ (٣٤)

Surah Maryam

Guidance

God created the Prophet Yahya and the Prophet Jesus in this manner as a sign to guide man (19:12,30,33). God is The Guide (*Al-Hādi* الهادى). The rest of the prophets continued the message on Earth (19:43,50,52,54–58). God confirms His guidance to those who are already on the right way (19:76).

God Does Not Need a Son

God is Absolute (*Al-Samad* الصمد) and He will punish, severely, anyone who disbelieves in Him or take partners with Him. However, He is The Just (*Al-'Adl* العدل), The Indulgent (*Al Afuww* العفو), and The Patient (*Al Sabūr* الصبور) (19:37-40,47,48,60-61,65,68-71,72-75,84–89)

How could God possibly need a son? He is The Great *(Al-'Aẓīm* العظيم*)*, The Mighty *(Al-'Azīz* العزيز*)*, The High *(Al-'Alīyy* العلى*)*, and The Bigger Than All *(Al-Kabīr* الكبير*)*. Whatever He wants He has only to say "Be!" and it is (19:35). The whole universe is shattered with such a claim (19:88–90). The heavens all but rip apart, the Earth splits asunder, the mountains crash from such a criminal claim that the All-Merciful has a son (19:91–92). He does not need sons like the Prophet Zachariah or the Prophet Abraham. All who are in the heavens and earth are but His subjects and will return to Him on their own on the Day of Doom.

Chapter Nine

The Equitable (Al-Muqsiṭ)

This chapter describes the natural phenomena that God used for selective, mass destruction. He mentions these signs to warn and advise the nation of Muhammad. The punishments came at the end of a long period of guidance through the prophets (discussed earlier in Chapter Five). *Surah*s 20 and 28 give the Prophet Moses as an example. They illustrate how prophets are selected, educated, qualified, supported with signs, then given the task of passing God's light to people through their message and holy books.

The Symbolic Letters: Ṭā Sīn Mīm (طسم) and Ṭā Sīn (طس)

The letters *ṭā*, *sīn*, and *mīm* form verse one of *surah*s 26 and 28 and the letters *ṭā* and *sīn* form the beginning of verse one of *surah* 27. Directly following these letters the Qur'an states that these are the signs of the manifest book. In *surah* 27, the *mīm* is missing (27:1) and the word "Qur'an" appears to confirm the identity of the book.

Computer and Arabic Dictionary Search
A computer search of the Qur'an shows that no words begin with these three letters. The Arabic dictionary states that they are only present in the three *surah*s in the Qur'an.

Explanation of the Symbolic Letters Ṭā Sīn Mīm (طسم) in the Light of the Beautiful Names of God
The letters *ṭā*, *sīn*, and *mīm* are included individually in forty-three of the names of God. Only one of the names contains all three letters, *Al-Muqsit* (المقسط), The Equitable. The three *surah*s complete one another. They have one theme: God is Equitable and Just. God does not destroy a

nation before He has warned its people and sent them books and signs (26:208–209). His system of guidance has been discussed earlier and is based on sending a messenger as a prophet (28:59). His rewarding system is in favor of man (28:84).

Never did We destroy a population, but had its warners — by way of reminder; and We never are unjust. (26:208–209)	وَمَا أَهْلَكْنَا مِنْ قَرْيَةٍ إِلاَّ لَهَا مُنْذِرُونَ(٢٠٨)ذِكْرَى وَمَا كُنَّا ظَالِمِينَ (٢٠٩) Surat al-Shu'arā'
Nor was thy Lord the one to destroy a population until He had sent to its center an messenger, rehearsing to them Our signs; nor are We going to destroy a population except when its members practise iniquity. (28:59)	وَمَا كَانَ رَبُّكَ مُهْلِكَ الْقُرَى حَتَّى يَبْعَثَ فِي أُمِّهَا رَسُولاً يَتْلُو عَلَيْهِمْ آيَاتِنَا وَمَا كُنَّا مُهْلِكِي الْقُرَى إِلاَّ وَأَهْلُهَا ظَالِمُونَ (٥٩) Surat al-Qaṣaṣ

Relationship among the Three *Surah*s

The beginning of the three *surah*s (26, 27, and 28) states that these are the signs of the Qur'an. These *surah*s emphasize the universe's natural phenomena and the danger that they may produce. The scientific recognition of these phenomena has been a long time coming. We have begun to recognize the possible effects of objects in space on Earth and their possible danger. The *surah*s also contain many scientific references related to the universe and to Earth in particular. When we recognize them scientifically we should be in a position to acknowledge God as the Creator (27:93).

And say: "Praise be to Allah, who will soon show you His signs, so that you shall know them"; and thy Lord is not unmindful of all that you do. (27:93)	وَقُلِ الْحَمْدُ لِلَّهِ سَيُرِيكُمْ آيَاتِهِ فَتَعْرِفُونَهَا وَمَا رَبُّكَ بِغَافِلٍ عَمَّا تَعْمَلُونَ (٩٣) Surat al-Naml

Possible Relation between the Three *Surah*s and Other *Surah*s

The Qur'an is a closely interwoven network of verses and *surah*s, some of which reflect parts of books sent to the people of Israel (26:196–197). It is also part of the Mother Book (3:7). The *surah*s narrate facts concerning some of the prophets, some of whom have their own books and probably the meaning of the verses in all these books is identical. The Prophet Moses is mentioned 129 times in 124 verses in 32 *surah*s in the Qur'an. Storing such information in the Mother Book can only be a task for the Creator. It will be shown later that the number of books that could possibly be kept by God are endless. No number of human machines, computers or compact disks could collate, classify, file, store, and retrieve such volumes of intricate information.

Without doubt it is (announced) in the mystic books of former peoples. Is it not a sign to them that the learned of the children of Israel knew it (as true)? (26:196–197)

وَإِنَّهُ لَفِي زُبُرِ الْأَوَّلِينَ(١٩٦)أَوَلَمْ يَكُنْ لَهُمْ آيَةً أَنْ يَعْلَمَهُ عُلَمَاءُ بَنِي إِسْرَائِيلَ (١٩٧)

Surat al-Shu'arā'

He it is who has sent down to you the Book: In it are verses basic or fundamental (of established meaning); they are the foundation of the Book: others are allegorical. But those in whose hearts is perversity follow the part thereof that is allegorical, seeking discord, and searching for its hidden meanings, but no one knows its hidden meanings except Allah. And those who are firmly grounded in knowledge say: "We believe in the Book; the whole of it is from our Lord" and none will grasp the Message except men of understanding. (3:7)

هُوَ الَّذِي أَنْزَلَ عَلَيْكَ الْكِتَابَ مِنْهُ آيَاتٌ مُحْكَمَاتٌ هُنَّ أُمُّ الْكِتَابِ وَأُخَرُ مُتَشَابِهَاتٌ فَأَمَّا الَّذِينَ فِي قُلُوبِهِمْ زَيْغٌ فَيَتَّبِعُونَ مَا تَشَابَهَ مِنْهُ ابْتِغَاءَ الْفِتْنَةِ وَابْتِغَاءَ تَأْوِيلِهِ وَمَا يَعْلَمُ تَأْوِيلَهُ إِلاَّ اللَّهُ وَالرَّاسِخُونَ فِي الْعِلْمِ يَقُولُونَ آمَنَّا بِهِ كُلٌّ مِنْ عِنْدِ رَبِّنَا وَمَا يَذَّكَّرُ إِلاَّ أُولُو الْأَلْبَابِ (٧)

Surat Al 'Imrān

The letter *ṭā* (ط) is present in the three *surah*s that refer to the Prophet Moses. It is also one of the symbolic letters of *surah* 20, *Ṭā Hā*. In both *surah*s 20 and 28, God names *Ṭūr* as the holy mountain on which God spoke to Moses — an important place at which God offers His guidance to man. The letter *ṭā* and the word *Ṭūr* refer to *surah* 52, which is titled *Al-Ṭūr*. The *surah* begins "*Al-Ṭūr* and the book *masṭūr*." *Masṭūr* means "written in lines," referring to the Qur'an; and it occurs in other verses in different *surah*s. *Ayah* 68:1 states: *Nun wa al-qalam wa ma yasturun* (*Nūn* and what the pen can write in lines). Thus the Qur'an indicates that the book mentioned in *ayat* 52:2 is the Qur'an and that the *nūn* will be used to rhyme 50.08 percent of the verses of the Qur'an. It is interesting to note that 84.6 percent, 90.32 percent, and 92.05 percent of the verses of *surah*s 26, 27 and 28 rhyme respectively, with the letter *nūn*.

Nūn, by the pen and by what it writes in lines and lines. (68:1) Surat al-Qalam	ن وَالْقَلَمِ وَمَا يَسْطُرُونَ (١)
By the mount (of revelation); by a decree inscribed in a scroll unfolded. (52:1–3) Surat al-Ṭūr	وَالطُّورِ(١)وَكِتَابٍ مَسْطُورٍ(٢)فِي رَقٍّ مَنْشُورٍ (٣)

Ṭā, sīn, and *mīm* (طسم) are the first three letters of the word *masṭūr* which means "in lines," but in the reverse order, i.e., as a mirror image. *Mīm* begins many names of God, including *Al-Musawwir,* The Imager. The first picture in imaging, that is the negative, is always reversed. When we look at any object, the picture that falls on the retina is reversed. It is put right in the brain in the occipital lobe. Reversing *ta sīn mīm* to read *mīm sīn ta* (مسط) yields the correct beginning of the word *masṭūr,* which is mentioned in *ayats* 52:2 and 68:1. If one adds the letter *qāf* (ق) after the *mīm* it gives the name of God, *Al-Muqsit* مقسط. The name is the theme of the three *surah*s.

In summary, the presence of the letter *ṭā* connects the four *surah*s (20, 28, 52, and 68). The *ṭā* refers to the sacred mountain *Ṭūr*. *Ṭūr* connects *surah*s 20 and 28 to illustrate the system of prophet guidance. It also points to *surah* 52 to show that the Qur'an is written in lines as referred to in *surah* 68. The latter gives the symbolic letter *nūn* which rhymes more than half of the verses of the Qur'an. A linguistic miracle. All this is controlled by the symbolic letters.

The above illustration suggests that the symbolic letters may be serving as markers for the names of God, *surah*s, prophets, holy places, or important events. Such letters may act like computer markers to relate and capture all this information and help to identify them into *umm al-kitāb* (the Mother Book).

Historical Signs of the Universe's Natural Phenomena

Table 11 lists the prophets preceding the Prophet Muhammad, why they were sent, and to whom, their signs, their message, and the universe's natural phenomena chosen for punishment. Their scientific explanation will be discussed in the next chapter.

Table 11. Prophets and Universe's Natural Phenomena

Name of Prophet	Signs	Crime	Natural Phenomena Punishment
Moses	9 signs including transform stick into snake, radiant hand, split the water. (26:15,32,33,45, 64; 27:12)	Pharaoh claimed he was a god, killed the male children of the Children of Israel and spared their women (28:4,39)	Split the water and drowning (26:64,67)
Abraham	God's attributes (26:75-87). God brings the sun from East (2:258). Fire is cool and safe for Abraham (21:69; 29:24).	Worshipping idols (26:71-72). Rejected signs & followed transgressors (11:59).	None. Fire in the life to come (26:91).
Noah	*Far al-tanūr* (11:40; 23:27; 69:11). The ark & the flood (26:119,120; 54:11-15).	Disbelieved (26:105).	Flood (26:22).
Hūd sent to the people of 'Ād	Bestowed on them knowledge, cattle, sons, gardens, & springs (26:132-135).	Worshipped idols. Built on high places to abuse passersby. Built underwater storage tanks as if they were immortal (26:128-129).	Lightening bolt (41:13). Hurricane-like wind (41:16,24,25; 51:41,42; 54:19, 20; 69:6-8)
Salih sent to the people of Thamūd	She camel free to eat & drink from God's earth in time of scarce water & pasture (26:155,156; 11:64; 50:27-29).	Worshipping idols, corrupting God's Earth (26:151,152). Hamstrung the camel (11:64, 68).	Blast (11:67; 54:31). Earthquake (7:78). Lightening bolt (41:13; 51:44).
Lot	Abraham told of sons Isaac & Jacob (11:71). His wife said they were too old (11:72).	Homosexuality (26:165,166; 27:55,56).	Brimstone (11: 82,83; 26:172, 173; 27:58; 54: 34). Rain (26: 173; 27:58). Blast (15:73).

Name of Prophet	Signs	Crime	Natural Phenomena Punishment
Shu'ayb sent to the Companions of the Wood	Unspecified sign (7:85; 11:88).	Worshipped idols. Corruption of the earth. Cheating, especially in measuring (26: 181-183).	Day of Shadow (26:189). Blast (11:9429:37). Earthquake (7:91; 29:37).
Solomon sent to his people and to Saba	Understood language of birds (27:16,20-24,27) and ants (27:18–19). Brought queen's throne (27:40-42). Glass floor of the palace (27:44).	Worshipped the Sun (27:24)	None.

Signs and Science

God states in the *surah*s under study that they contain signs or miracles. We have enumerated the scientific natural phenomena of the universe that God used for precise, mass destruction. Other sciences are also mentioned, examples of which are given below.

(Here is) a book which We have sent down unto you, full of blessings, that they may meditate on its signs, and that men of understanding may receive admonition. (38:29)

كِتَابٌ أَنزَلْنَاهُ إِلَيْكَ مُبَارَكٌ لِيَدَّبَّرُوا آيَاتِهِ وَلِيَتَذَكَّرَ أُولُو الْأَلْبَابِ (٢٩)

Surah Ṣād

Social Sciences

The Qur'an, which is written in a symbolic, miraculous Arabic language (26:195), is mentioned in the preceding holy books (26:196). Part of it is recognized by the knowledgeable people of Israel (26:197). It is made for our guidance (27:92 and 16:64). We are ordered to reflect on its verses and signs (38:29).

The Qur'an contains a wealth of information on social sciences. It introduces systems of taxation, contracts, testimonials, wills, civil, and criminal law.

Physical Sciences

God has given us scientific references to help our progress in science. He is patient with us during our perseverance to attain knowledge (17:44); however, when we reach our scientific targets and unravel the secrets of His scientific allusions we shall recognize Him as the Creator (27:93).

Science References—More Befitting Signs in Our Time Than Sticks That Transform into Snakes

Although the followers of the Prophet Muhammad believed in him and his message they repeatedly asked him why he could not have an *ayah* from God to substantiate his authenticity. The word "ayah" in Arabic could mean verse, miracle, proof, evidence, lesson, revelation, etc. To understand how Allah satisfied the peoples request one has to study this question in the Qur'an by studying the *surah*s in their cronological order of revelation. At first, people requested the Prophet Muhammad to show them an *ayah* like those given to the Prophet Musa (28:48) (*surah* 49, cronologically). God replied in the same verse that the signs that He gave to the Prophet Musa were disbelieved by his people and he and his brother were considered magicians. The first sign he was given was the conversion of his stick into a snake which swallowed all that the magicians of Pharaoh could produce in magic. In *Surah Yunus* (10:20; *surah* 51, cronologically) again the followers of the Prophet Muhammad repeatedly requested a sign from Allah to him, God replied that only He is aware of the future and that they should wait and so should the Prophet (as the signs are a matter for the future). A little later, in *Surah Yusuf* (12:105; *surah* 53, cronologically), God informed them that He had given them many signs in the heavens and the earth but they ignored them and they could not appreciate their significance. Such signs like the eclipse of the sun or moon, falling stars, earthquakes, variation of the day and the night, stages of the moon, etc., all meant nothing to them. The answer to the problem came in *Surah al-An'am* (*surah* 6; 55 cronologically). God states the repeated requests by people for an *ayah*, He reveals to them that science references are one of many miracles He gave them in the Qur'an and even after He states such science references to them, some of them will still deny. In verses 4, 25, and 37 of *surah* 6, God states that the people turned away from all the signs He gave them (6:4) and that on some of them He sets veils on their hearts and deafness in their ears so that they cannot believe in His signs (6:25). However, they still asked for a sign from the Prophet Muhammad (6:37). Allah's reply came in verse 6:38 in which He states that moving creatures on earth and birds that fly with two wings are nations like man. This great scientific reference will remain a scientific challenge to man. So far we understand very little about the language of animals, including birds. We understand the communicaton between dolphins and we know that bees can communicate by "dancing" movements. Extensive research is now going on to understand the language of animals and birds for the benefit of man. It is very unlikely that we will ever, even in thousands of years, understand the language of all these creatures. Such a science reference and many others in the Qur'an are the signs suitable for our time that Allah has given us in the Qur'an to boast about and speak of to other nations with different beliefs. In the next verse (6:39), even when all such signs are shown some people remain in darkness, deaf and mute. People have vowed that when they receive a sign from God they will believe in it (6:109) and God says that the signs are with Him and that when people receive it they will still not believe in them. Allah in His wisdom, aware of the future, knew that He will bestow on Muhammad's nation the blessing of knowledge and science. Instead of giving him signs like Musa, a stick that transforms into a snake, He put science references into the Qur'an that could only be understood after thousands of years after the revelation of the Qur'an. These are some of the miracles of the Qur'an and for this reason *Science Miracles: No Sticks or Snakes* was chosen as a title for this book.

But (now), when the Truth has come to them from Ourselves, they say: "Why are not (signs) sent to him, like those which were sent to Musa? Do they not then reject (the signs) which were formerly sent to Musa? They say: "Two kinds of sorcery each assisting the other!" And they say: "For us, we reject all (such things)!" (28:48)

فَلَمَّا جَاءَهُمُ الْحَقُّ مِنْ عِنْدِنَا قَالُوا لَوْلَا أُوتِيَ مِثْلَ مَا أُوتِيَ مُوسَى أَوَلَمْ يَكْفُرُوا بِمَا أُوتِيَ مُوسَى مِنْ قَبْلُ قَالُوا سِحْرَانِ تَظَاهَرَا وَقَالُوا إِنَّا بِكُلٍّ كَافِرُونَ (٤٨)

Surat al-Qaṣaṣ

And they say: "How is it that not a sign is sent down on him from his Lord?" Say: "The unseen belongs to Allah alone, so wait you, verily. I am with you among those who wait (for Allah's judgement)." (10:20)

وَيَقُولُونَ لَوْلَا أُنْزِلَ عَلَيْهِ آيَةٌ مِنْ رَبِّهِ فَقُلْ إِنَّمَا الْغَيْبُ لِلَّهِ فَانْتَظِرُوا إِنِّي مَعَكُمْ مِنَ الْمُنْتَظِرِينَ (٢٠)

Yūnus

And how many a sign in the heavens and the earth they pass by, while they are averse therefrom. (12:105)

وَكَأَيِّنْ مِنْ آيَةٍ فِي السَّمَاوَاتِ وَالْأَرْضِ يَمُرُّونَ عَلَيْهَا وَهُمْ عَنْهَا مُعْرِضُونَ (١٠٥)

Yūsuf

And never an *ayah* comes to them from the *ayāt* (proofs, evidences, verses, lessons, signs, revelations) of their Lord, but that they have been turning away from it. (6:4)

وَمَا تَأْتِيهِمْ مِنْ آيَةٍ مِنْ آيَاتِ رَبِّهِمْ إِلَّا كَانُوا عَنْهَا مُعْرِضِينَ (٤)

Surat al-Anʿām

And of them there are some who listen to you; but We have set veils on their hearts, so they understand it not, and deafness in their ears; and even if tey see every one of the *ayāt* they will not believe therein; to the point that when they come to you to argue with you, the disbelievers say: "These are nothing but tales of the men of old." (6:25)

وَمِنْهُمْ مَنْ يَسْتَمِعُ إِلَيْكَ وَجَعَلْنَا عَلَى قُلُوبِهِمْ أَكِنَّةً أَنْ يَفْقَهُوهُ وَفِي آذَانِهِمْ وَقْرًا وَإِنْ يَرَوْا كُلَّ آيَةٍ لَا يُؤْمِنُوا بِهَا حَتَّى إِذَا جَاءُوكَ يُجَادِلُونَكَ يَقُولُ الَّذِينَ كَفَرُوا إِنْ هَذَا إِلَّا أَسَاطِيرُ الْأَوَّلِينَ (٢٥)

Surat al-An'ām

They say: "Why is not a sign sent down to him from his Lord?" Say: "Allah has certainly power to send down a sign: But most of them understand not." There is not an animal (that lives) on the earth, nor a being that flies on its wings, but (forms part of) communities like you. Nothing have we omitted from the Book, and they (all) shall be gathered to their Lord in the end. Those who reject Our *ayāt* are deaf and dumb in the darness. Allah sends astray whom He wills and He guides on the straight path whom He wills. (6:37–39)

وَقَالُوا لَوْلَا نُزِّلَ عَلَيْهِ آيَةٌ مِنْ رَبِّهِ قُلْ إِنَّ اللَّهَ قَادِرٌ عَلَى أَنْ يُنَزِّلَ آيَةً وَلَكِنَّ أَكْثَرَهُمْ لَا يَعْلَمُونَ (٣٧) وَمَا مِنْ دَابَّةٍ فِي الْأَرْضِ وَلَا طَائِرٍ يَطِيرُ بِجَنَاحَيْهِ إِلَّا أُمَمٌ أَمْثَالُكُمْ مَا فَرَّطْنَا فِي الْكِتَابِ مِنْ شَيْءٍ ثُمَّ إِلَى رَبِّهِمْ يُحْشَرُونَ (٣٨) وَالَّذِينَ كَذَّبُوا بِآيَاتِنَا صُمٌّ وَبُكْمٌ فِي الظُّلُمَاتِ مَنْ يَشَأْ اللَّهُ يُضْلِلْهُ وَمَنْ يَشَأْ يَجْعَلْهُ عَلَى صِرَاطٍ مُسْتَقِيمٍ (٣٩)

Surat al-An'ām

And they swear their strongest oaths by Allah that if there came to them a sign they would surely believe therein. Say: "Signs are but with Allah and what will make you (Muslims) perceive that (even) if it (the sign) came, they will not believe?" (6:109)

وَأَقْسَمُوا بِاللَّهِ جَهْدَ أَيْمَانِهِمْ لَئِنْ جَاءَتْهُمْ آيَةٌ لَيُؤْمِنُنَّ بِهَا قُلْ إِنَّمَا الْآيَاتُ عِنْدَ اللَّهِ وَمَا يُشْعِرُكُمْ أَنَّهَا إِذَا جَاءَتْ لَا يُؤْمِنُونَ (١٠٩)

Surat al-An'ām

God created the heavens and earth and He spreads His authority on them (27:26,60). He knows what is hidden in them; He is aware of our thoughts and the future (27:25,65,74–75).

He made the earth for man to settle and He sent rainwater for his existence, for plants and animals (27:60). He made rivers and created the *rawasi* or lithospheres which carry our continents and extend the earth at the oceanic ridges (15:19; 50:7) and reduces it at the trenches (13:41). He created the estuary cycle as a barrier between rivers and seas (27:61).

He guided us in the darkness of land and sea even before the compass was known. He arranged the stars into constellations (25:61) and staged the moon in houses (36:39) as markers during the night. He could have made us face the sun all of the time, giving half the Earth continuous day and half continuous night (28:71–73). He sent the winds at special levels of latitude to help our navigation and existence (27:63). He appointed us successors on Earth to discover some of His scientific signs (27:62).

> Or, who listens to the (soul) distressed when it calls on Him, and who relieves its suffering, and makes you (mankind) inheritors of the earth? (Can there be another) god besides Allah? Little it is that you heed! (27:62)
>
> أَمَّنْ يُجِيبُ الْمُضْطَرَّ إِذَا دَعَاهُ وَيَكْشِفُ السُّوءَ وَيَجْعَلُكُمْ خُلَفَاءَ الْأَرْضِ أَإِلَهٌ مَعَ اللَّهِ قَلِيلًا مَا تَذَكَّرُونَ (٦٢)
>
> Surat al-Naml

Planned Universe's Natural Phenomena

Disbelieving people wondered when their punishment would be, challenging Him to hasten it (27:71). God hinted that it might have already started (27:72). Perhaps He is referring to the universe's natural phenomena that He used repeatedly for mass punishment, of which examples are numerous. According to our present knowledge, the Earth is under threat from many asteroids, one of which is Swift-Tuttle. It will reach us in about 30 years. It will come near the Earth and will miss our orbit and atmosphere by about 14 days. It is 15 miles in diameter and if it collided with the Earth, life would cease to exist.

> They also say: "When will this promise (come to pass)? (Say) if you are truthful." Say: "It may be that some of the events which you wish to hasten on may be (close) in your pursuit!" (27:71–72)
>
> وَيَقُولُونَ مَتَى هَذَا الْوَعْدُ إِنْ كُنْتُمْ صَادِقِينَ (٧١) قُلْ عَسَى أَنْ يَكُونَ رَدِفَ لَكُمْ بَعْضُ الَّذِي تَسْتَعْجِلُونَ (٧٢)
>
> Surat al-Naml

CHAPTER TEN

THE UNIVERSE'S NATURAL PHENOMENA

God used the universe's natural phenomena for select, precise mass destruction. He spared the innocent with a merciful system, which we will discuss in Chapter Eleven. The Qur'anic description of these phenomena is sometimes so accurate that one might be lead to assume that he is reading a book of science; however, the scientific facts are given piecemeal. Since the Prophet Muhammad could neither read nor write and the people who received his message were also illiterate, the scientific information was given to them in small doses and simplified.

In this chapter an attempt is made to give a possible scientific explanation of these phenomena; however, their secrets will remain with God. Science has taken great strides in the explanation of many of the universe's natural phenomena. We shall use current knowledge to explain how the events took place. It is very important to appreciate what has happened in the past, as it may recur in the future. Man, being God's successor on earth, has to live up to expectations. He has to defend himself, wherever possible, from earthquakes, floods, hurricanes, and possible invasion of Earth by objects from outer space.

The Flood of Noah: The Qur'anic Description

Preparation for the Flood

Noah lived 950 years (29:14) to give him time to build his ark, which was made of fastened planks (54:13). Its manufacture required skill, thus God supported him with vision and thought (23:27), and during the flood He directed him at sea (54:14).

We do not know when exactly Noah lived. During his long life many changes could have taken place on Earth. He might have lived in the era of melting ice caps, impacts on Earth by meteorites and asteroids or a tsunami which may have caused flooding and death on land.

We (once) sent Noah to his people, and he tarried among them a thousand years less fifty: but the deluge overwhelmed them while they (persisted in) sin. (29:14)	وَلَقَدْ أَرْسَلْنَا نُوحًا إِلَى قَوْمِهِ فَلَبِثَ فِيهِمْ أَلْفَ سَنَةٍ إِلاَّ خَمْسِينَ عَامًا فَأَخَذَهُمُ الطُّوفَانُ وَهُمْ ظَالِمُونَ (١٤) Surat al-'Ankabūt
So We inspired him (with this message): "Construct the Ark within Our sight and under Our guidance: then when comes Our command, and the fountains of the earth gush forth, take on board pairs of every species, male and female, and your family — except those of them against whom the Word has already gone forth: and address Me not in favor of the wrongdoers; for they shall be drowned (in the Flood). (23:27)	فَأَوْحَيْنَا إِلَيْهِ أَنِ اصْنَعِ الْفُلْكَ بِأَعْيُنِنَا وَوَحْيِنَا فَإِذَا جَاءَ أَمْرُنَا وَفَارَ التَّنُّورُ فَاسْلُكْ فِيهَا مِن كُلٍّ زَوْجَيْنِ اثْنَيْنِ وَأَهْلَكَ إِلاَّ مَن سَبَقَ عَلَيْهِ الْقَوْلُ مِنْهُمْ وَلاَ تُخَاطِبْنِي فِي الَّذِينَ ظَلَمُوا إِنَّهُم مُّغْرَقُونَ (٢٧) Surat al-Mu'minūn

Description of the Flood

The flood was initiated by what God describes as *far al tanur* (11:40; 23:27). *Far* means "overflow" and *al-tanur* means "face of the earth" as described by the Prophet Muhammad's nephew Ali ibn Abu Talib. Heaven's gates opened with torrential water and the earth burst with springs, and both waters met at a fated decree (54:11–12). The sea level rose to an exceptionally high level (69:11), giving rise to mountainous waves (11:42).

At length, behold! There came Our command, and the fountains of the earth gushed forth! We said: "Embark therein, of each kind two, male and female, and your family — except those against whom the word has already gone forth —	حَتَّى إِذَا جَاءَ أَمْرُنَا وَفَارَ التَّنُّورُ قُلْنَا احْمِلْ فِيهَا مِن كُلٍّ زَوْجَيْنِ اثْنَيْنِ وَأَهْلَكَ إِلاَّ مَن سَبَقَ عَلَيْهِ الْقَوْلُ وَمَنْ آمَنَ وَمَا آمَنَ مَعَهُ إِلاَّ قَلِيلٌ (٤٠) Surah Hūd

and the believers. But only a few believed with him. (11:40)

We, when the water (of Noah's Flood) overflowed beyond its limits, carried you (mankind), in the floating (Ark). (69:11)

إِنَّا لَمَّا طَغَى الْمَاءُ حَمَلْنَاكُمْ فِي الْجَارِيَةِ (١١)

Surat al-Ḥāqqah

So the Ark floated with them on the waves (towering) like mountains, and Noah called out to his son, who had separated himself (from the rest): "O my son! embark with us, and be not with the unbelievers!" (11:42)

وَهِيَ تَجْرِي بِهِمْ فِي مَوْجٍ كَالْجِبَالِ وَنَادَى نُوحٌ ابْنَهُ وَكَانَ فِي مَعْزِلٍ يَابُنَيَّ ارْكَبْ مَعَنَا وَلاَ تَكُنْ مَعَ الْكَافِرِينَ (٤٢)

Surah Hūd

Possible Scientific Explanations

Melting of the Ice Caps

Noah may have lived through a time of major changes in the Earth's topography. Twenty thousand years ago there was no North Sea and 8,000 years ago we could walk from England to France. We are now living in an ice age that might extend for 2,000,000 years. Scientists define an ice age by the presence of ice caps. Many theories have been given to explain the time and reasons for the building or retreating of the ice caps (pulse beat of the ice age and the theory of solar radiation). Some of the most accepted theories are given below.

The Mathematical Theory

Milutin Milankovitch (Figure 5), a Yugoslavian geophysicist, proposed his astronomical theory of climate change in 1930. He mathematically predicted how the ice sheets grow and retreat at the poles. This would depend on how much of the sun's radiation hits the poles. According to his theory the amount of radiation reaching the poles is dependent on the tilt of the Earth towards the sun and how near or far the Earth would be as it orbits the sun. As the Earth orbits the sun in an ellipse it is sometimes closer and sometimes farther away from the sun. Milankovitch speculated that this difference varies with time and is maximized at 10,000,000 km every 100,000 years. As the Earth orbits the sun it tilts on its axis of rotation in the plane of the orbit. The tilt varies with time from less than 22 degrees to a little more than 24 degrees, but changes very regularly for a period of 40,000 years. As the Earth is spinning it wobbles with a period of 22,000 years (Figure 5). All this affects the amount of solar radiation falling on the poles.

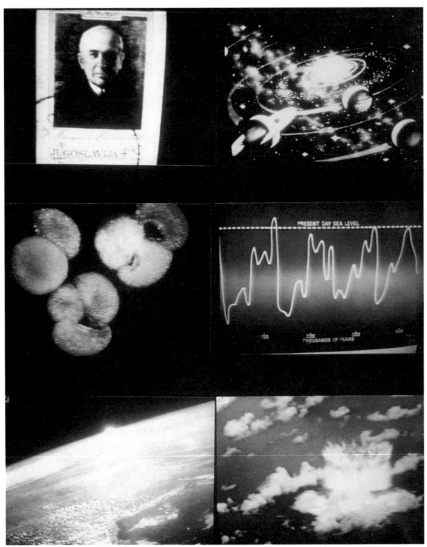

Figure 5: *(TL)* Milutin Milankovitch 1879–1958. *(TR)* Earth going around the Sun in an elliptical orbit, with eccentricity, inclination and precession that vary at regular periods of 100,000, 40,000, and 22,000 years respectively. *(ML)* Foraminifera. *(MR)* Present day sea level, light red and heavy blue oxygen molecules in ocean water, peaks of periodic ice cap build up. It corresponds to the geological, Earth's magnetic field, and Milankovitch theories. *(BL)* Asteroid impact on Earth. *(BR)* Asteroid impact in the ocean.

During cold summers no ice will melt and so the ice caps will not retreat.

Geological Studies

The Milankovitch theory has been supported by studies of deep-sea fossils. Crustaceans by the name of Foraminifera (Figure 5) use the dissolved oxygen in the water of the oceans in the construction of their shells. The dissolved oxygen in the ocean is present in both heavy and light molecules. During evaporation, clouds are rich in the light oxygen water atoms. The heavy oxygen water stays in the sea. The rain returns light oxygen molecules back to the sea. In the ice age, the rain of light oxygen vapor drops as snow on the ice cap. When there is an ice cap the seawater will have proportionally more heavy oxygen water molecules. Foraminiferas record the type of ocean water molecules that it absorbs to make its shell. In other words, they act as a tape recorder of what was present in the ocean, giving the history of the ice caps.

The curve in Figure 5 shows the proportion of light (red) and heavy (blue) oxygen molecules in the ocean water. Sediments falling to the seabed align themselves in the direction of the Earth's magnetic field. The Earth's magnetic field flips into reverse direction at different intervals of time. It flipped 700,000 years ago. In 1976, three scientist — James Hays, John Imbrie, and Nick Shackleton — put time on these variations of light and heavy oxygen water molecules into the curve. These are the peaks that appear in the graph at 22,000, 40,000, and 100,000 years (Figure 5).

Sea level variations are shown to correspond to the concentration of heavy and light oxygen molecules in the water. Now it is at its highest, it could be 120 meters less as shown in the graph. Ice reflects the Sun's radiation so the Earth cools and more ice is formed. When ice retreats, less of this mirror is present, more of the Sun's radiation is absorbed by the Earth, thus further encouraging snow to melt and the ice cap to retreat.

Meteorite Impact

Geologists believe that a collision by one of the universe's objects, probably an asteroid, with the Earth 2.3 million years ago, was implicated with an ice age. The collision of a meteorite the size of a football with the Earth's atmosphere could cause an explosion equal in power to 4 or 5 times that of the atomic bombs dropped on Hiroshima and Nagasaki. A meteorite of a larger size colliding with the Earth on the ice caps or in the oceans could cause a flood of the magnitude of that described in the Qur'anic version of Noah's flood (Figure 5).

66   Science Miracles: No Stick or Snakes

Figure 6: *(TL & TR) Tsunamis. (BR) Tsunami detected on the surface of the ocean. It receives signals from a sensor at the bottom of the ocean (BL) of any earthquake. It sends the signals to satellites (MR). The satellites send the signals (ML) to earth stations (**Fig.7** MR).*

Tsunami or Seismic Sea Waves

These are the most catastrophic of all ocean waves (Figure 6). They are generated by tectonic displacements as a result of volcanoes, landslides, or earthquakes on the sea floor. These will cause a sudden displacement of the water directly above them and the formation of a group of water waves possessing wavelengths equal to the water depth, which could be several thousand meters. These waves can travel radially outwards for thousands of kilometers while retaining substantial energy (Figure 7). Their speed is generally about 500 km/hr (300 mph). In the open ocean their amplitude is usually less than one meter, therefore, they usually pass unnoticed by ships

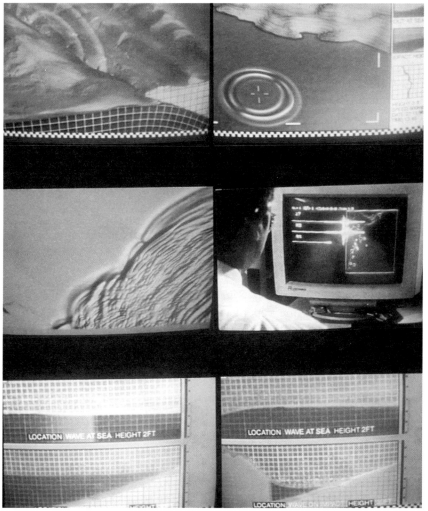

Figure 7: *(TL) Earthquake between two tectonic plates. (TR) It creates a wave that travels (ML) along the bottom of the ocean. It moves along the floor of the ocean (BL & BR) and appears at the coast as a tsunami. It can be detected by the warning system in the Pacific (MR).*

at sea. However, as they approach a beach, the resultant waves can be devastating, especially to low-lying coastal areas. The 37 m (120 ft) waves from the 1883 Krakatoa eruption, for example, killed 36,000 people. Earthquakes producing tsunamis usually exceed 6.5 on the Richter scale. Most tsunamis occur in the Pacific Ocean where a warning system has now been established (Figure 6).

The warning system consists of strategically placed seismic stations and a communications network (Figure 6). A new system, using sensors on the bottom of the ocean to sense seismic activity, sends its signals to stations floating on the ocean's surface. The signals from these stations are picked up by satellite, which in turn are sent to land-based stations in areas of potential danger (Figure 7).

Tsunamis are known to occur on shores of continents surrounded by oceans, however, they may occur elsewhere. The Mediterranean area has always been the site of earthquakes and instability. These may have triggered seismic waves that caused flooding and destruction. Many famous cities in this area have been destroyed by tsunamis. Alexandria was destroyed by a tsunami after an earthquake, which levelled it to the ground in 365 A.D. Tsunamis are known to reach as far as four and one-half km inland as in Peru on August 23, 1868.

Meteor, Meteorites, Asteroids and Comets

A **meteor**, known as a shooting or falling star (Figure 8), is the streak of light produced by the vaporization of interplanetary particles as they enter the Earth's atmosphere. A large and abnormally bright meteor is referred to as a fireball. Larger meteors are not completely vaporized, and the particles that reach the Earth's surface are called meteorites. They may form meteorite craters such as those seen on the Moon and the satellites of Mars and Jupiter. Meteorites falling on Jupiter have been well recorded by different observers on Earth (Figure 9). One of the craters produced there was as large as the planet Earth (Figure 9).

Meteors can be seen on any night after midnight. During certain times of the year, many are visible so that they are termed "meteor showers." They are believed to be produced by the debris of comets. They strike the Earth's atmosphere at speeds between 35 and 95 kps. The average meteor is about the size of a grain of sand.

Micrometeorites drift to the ground without vaporizing and may add 1,000 tons to the mass of the Earth each day. Although 500 large meteorites

Science Miracles: No Stick or Snakes

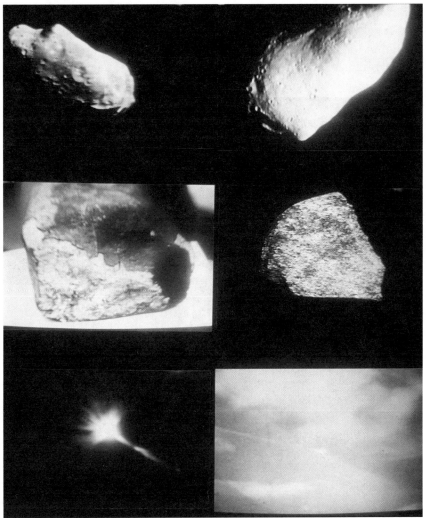

Figure 8: (TL) Asteroid 243 IDA more than twice the size of Gaspara. (TR) Asteroid 951 Gaspara, as large as the island of Manhattan. (ML) Iron meteorite. (MR) Meteorite that contains gas similar to atmosphere on Mars, found in Egypt. (BL) Halley's comet. (BR) Falling star.

may fall on the Earth each year, only about one percent of these are recovered.

Meteorites are classified according to their composition as irons (siderites), stony irons (siderolites) and stones (aerolites). Iron meteorites are the type most often found because of their appearance. Stone meteorites are more abundant and are easily mistaken for ordinary rocks.

70   Science Miracles: No Stick or Snakes

A special class of meteorites are the carbonaceous chondrites. These meteorites consist largely of the mineral serpentine, in addition, they contain amino acids and other organic compounds believed to be of extraterrestrial origin, but which most of the evidence points to a nonbiological ori-

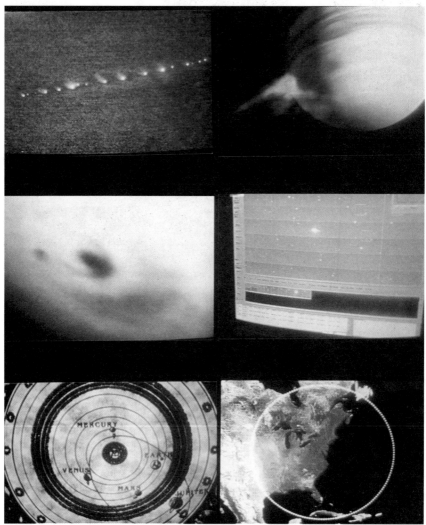

Figure 9: (TL) Comets heading to impact on Jupiter. (TR) Comet impact on Jupiter. (ML) Comet impact on Jupiter causing crater equal to the size of Earth. (MR) Earth-based space watch recording meteorites that reach Earth's orbit. (BL) Asteroid belt. (BR) Impact in the United States of a comet or asteroid of 5 miles in diameter may cause its total destruction.

gin. Nine meteorites named SNCs (shergottites-nakhlites-chassignites), referring to the localities where they were found (India, Egypt and France respectively), may have come from Mars (Figure 8). Mass spectrometric analysis of gas liberated from these meteorites indicates a composition similar to Mars's atmosphere analyzed by Viking Landers (Figure 8).

Asteroids are present in large numbers as small solid objects. Sometimes they are called minor planets because they orbit the Sun in a swarm called the Asteroid Belt, which is present between the orbits of Mars and Jupiter. The known asteroids include: about 75 Amor asteroids whose orbits intersect the orbit of Mars; about 50 Apollo asteroids whose orbits intersect the Earth's orbit and 16 Trojan asteroids which follow Jupiter in its orbit (Figures 8, 9, and 10). In 1989, one Apollo asteroid passed within 8,000 km of Earth.

Ceres, Pallas, and Vesta are the three largest asteroids. They have diameters of 785 km, 610 km, and 540 km, respectively. About half a million asteroids have diameters over 1 km and others could range down to the size of meteorites. Asteroids are probably bodies from the first days of the solar system. Collisions between them must have been numerous, as indicated by their irregular shape. It is estimated that three-quarters of the main-belt asteroids with diameters greater than 50 km have surfaces similar to carbonaceous chondrites.

Comets are celestial bodies of small mass. Usually, they travel around the Sun in an elongated orbit, becoming visible as they near the Sun and sometimes forming a tail. For centuries they were considered harbingers of catastrophes. Sir Isaac Newton was the first to discover the true orbit of a comet from its observed trajectory in the sky. Edmund Halley, who discovered Halley's Comet, showed that it had the same orbit in 1531, 1607, 1682, and predicted it would return in 1758, and it did. It has been observed 20 times since 239 AD. Its most recent appearance occurred in 1985–86 (Figure 8). Comets' orbits are elliptical. Short-period comets have periods of less than 200 years. The shortest known period for a comet is 3.3 years (Enck's comet). Some long-period comets may take several thousand years.

Almost the entire mass of a comet is concentrated in its nucleus, which is only a few kilometers in diameter. The density, 0.1–1 g/cu-cm, indicates it is very tenuous. Fred L. Whipple's "dirty snowball" model (Figure 1) shows the nucleus to consist of a mixture of compounds such as water, carbon dioxide, ammonia and methane all frozen and mixed with grit and dust. When the comet approaches the Sun, this frozen matter sublimes and forms a cloud of gas and grit, called a coma, around the nucleus. As the comet approaches the Sun the production of gases increases. The gas and dust particles are repelled from the nucleus by the solar radiation pressure and the solar wind (a stream of charged particles), forming the tail. The coma is

Figure 10: (**TL**) *Meteorite falls to Earth's atmosphere.* (**TR**) *Asteroid belt.* (**ML**) *Impact crater.* (**BL**) *The medieval cathedral of St George, Nordlingen, Southern Germany, built in an impact crater 24 km in diameter* (**BR**). (**MR**) *Stone wall of the cathedral containing black glass from shocked and melted rocks. Both changes are evidence of an impact. This find was the subject of a publication by Gene Shoemaker, comet discoverer, who defended the idea of impact craters on Earth.*

about 100,000 km. The tail is formed of gas from the coma and always points away from the Sun. Its length ranges from 1 km to 100 km.

Many comets, especially short-period ones, slowly disintegrate under the influence of the Sun's gravitational force. They may also collide with the Sun. They leave waste products behind in their orbits as meteoroids. When the Earth crosses such an orbit, meteor showers are frequently observed.

Scientists speculate that collisions of comets with the Earth may occasionally occur with devastating effects. In one hypothesis, a collision threw enough dust into the Earth's atmosphere to block the Sun and cause the extinction of some species of animals, like the dinosaur, and some plants (Figure 11). The tremendous blast in the Tunguska region of Russia, in June 1908, may be due to a similar impact.

In 1950 Jan Oort proposed that the Sun is surrounded by an enormous cloud of comet material at a distance of about 1,000 times that of the radius of the known Solar System (Figure 1). In 1951 Gerard Kuiper proposed that a ring of comet material lies in the plane of the Solar System several hundred times as far from the Sun as the Earth is. Both theories are accepted and both clouds merge according to astronomers. Perturbations by interstellar clouds or passing stars would cause some of the comet material to dislodge and enter the inner solar system in the form of comets. Short-period comets more likely arise from the Kuiper belt.

The brief description of objects from outer space that can collide with Earth shows a few of the many weapons that God has in His armory to cause a flood on Earth. God facilitated such events by prolonging Noah's life to almost 1,000 years. Melting of the ice caps by the Sun's radiation or by collision of a meteorite — a meteorite impact in the ocean or a tsunami — can result in a destructive flood.

The Blast That Turned Land Upside Down and the Rain with Labeled Stones from Hell

Another of the universe's natural phenomena has been described in the Qur'an on two occasions. Both are characterized by destruction with stones of *sijjīl*. The traditional explanation is that *sijjīl* means "baked in the fire of Hell." On one occasion the stones came down in the form of rain (11:82-83; 26:173; 27:58). On the second occasion, the stones of *sijjīl* were thrown down by some form of birds or "flyers" that flocked over the "companions of the elephant" (105:3–4).

We rained down on them a shower (of brimstone): And evil was the shower on those who were admonished (but heeded not)! (26:173)

وَأَمْطَرْنَا عَلَيْهِمْ مَطَرًا فَسَاءَ مَطَرُ الْمُنْذَرِينَ (١٧٣)

Surat al-Shu'arā'

And We rained down on them a shower (of brimstone): and evil was the shower on those who were admonished (but heeded not)! (27:58)

وَأَمْطَرْنَا عَلَيْهِمْ مَطَرًا فَسَاءَ مَطَرُ الْمُنْذَرِينَ (٥٨)

Surat al-Naml

The Qur'anic Description
Lot's People

Early in the morning (11:81), the people of Lot were awakened by a horrendous blast, or, as described in the Qur'an, by a great "cry." This was followed by their city being turned upside down (11:82). Then a rain of stones, baked in the fire of hell, fell on them layer upon layer (11:82). Each stone was labelled to hit a specific person (11:83).

(The messengers) said: "O Lut! We are messengers from your Lord! By no means shall they reach you! Now travel with your family while yet a part of the night remains, and let not any of you look back: but your wife (will remain behind): to her will happen what happens to the people. Morning is their time appointed: Is not the morning nigh?" When Our decree issued, We turned (the cities) upside down, and rained down on them brimstones hard as baked clay, spread, layer on layer — marked as from your Lord: Nor are they ever far from those who do wrong! (11:81–83)

قَالُوا يَالُوطُ إِنَّا رُسُلُ رَبِّكَ لَنْ يَصِلُوا إِلَيْكَ فَأَسْرِ بِأَهْلِكَ بِقِطْعٍ مِنَ اللَّيْلِ وَلَا يَلْتَفِتْ مِنْكُمْ أَحَدٌ إِلَّا امْرَأَتَكَ إِنَّهُ مُصِيبُهَا مَا أَصَابَهُمْ إِنَّ مَوْعِدَهُمُ الصُّبْحُ أَلَيْسَ الصُّبْحُ بِقَرِيبٍ (٨١) فَلَمَّا جَاءَ أَمْرُنَا جَعَلْنَا عَالِيَهَا سَافِلَهَا وَأَمْطَرْنَا عَلَيْهَا حِجَارَةً مِنْ سِجِّيلٍ مَنْضُودٍ (٨٢) مُسَوَّمَةً عِنْدَ رَبِّكَ وَمَا هِيَ مِنَ الظَّالِمِينَ بِبَعِيدٍ (٨٣)

Surah Hūd

Science Miracles: No Stick or Snakes

Companions of the Elephant

An army came with an elephant to destroy the Ka'ba but God sent flocks of birds or bird-like flying animals that threw on them stones baked in the fire of hell. By the time the attack was over, the place was empty and looked as if it were strewn with the dirty remains of stalks and straw eaten and stepped on by cattle (105:3–5).

And He sent against them flights of birds, striking them with stones of baked clay. Then did He make them like an empty field of stalks and straw, (of which the corn) has been eaten up. (105: 3–5)	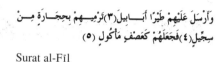 Surat al-Fīl

Possible Scientific Explanation

The Qur'anic description of the destruction of Lot's people portrays, more or less, the results from the impact of a meteorite, asteroid, or comet (Figure 11). Impacts of falling objects onto Earth, such as meteorites or asteroids, cause blasts. It would turn the site where it fell upside down. At the time of impact, earth would fly up into the air to a tremendous height but would come down again as stones and fire leaving no one alive. The impact crater which was recently discovered at Chicxulub on the Yucatan peninsula in the Gulf of Mexico, which has a diameter of 150 miles, occurred about 65 million years ago (Figure 11). Geologists give this as a reason for the extinction of many animals, including the dinosaurs, and plants. Other impact craters are present throughout the world. The medieval city of Nordlingen in southern Germany is built in an impact crater (Figure 10). Its 500-year-old St. George's cathedral is built from local stone that contains black glass, formed from shocked and melted rocksand; it also contains coecite,

Other scientific explanations can be given, such as volcanic eruption, tornadoes, etc., however, they cannot give as complete a description of the Qur'anic picture as the impact of a falling object from outer space onto Earth.

The stones dropped on the Companions of the Elephant, carried by flocks of flying objects or birds, suggest objects from outer space. It has been explained earlier that meteorites fall all the time on Earth without being detected. God might have chosen to send enough of them at one time to destroy the invaders.

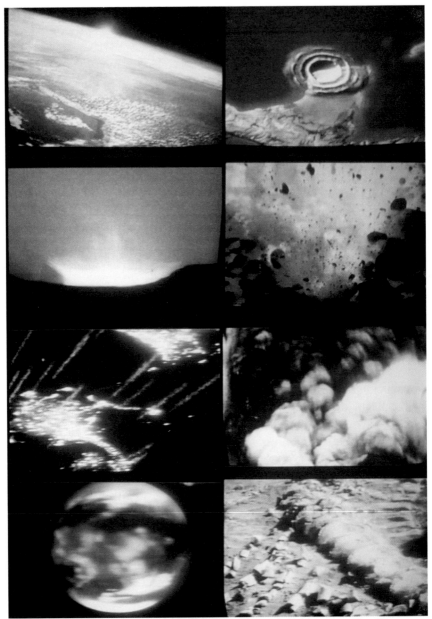

Figure 11: *(**TL**) Impact on Earth. (**TR**) Impact on the Gulf of Mexico 65 million years ago, 190 miles in diameter. (**MUL**) Impact on Earth. (**MUR**) Earth contents rocks, etc., flying in the air; they come back as fire, rocks (**LLM**), and smoke (**RLM**). (**BL**) Fires develop on Earth and the smoke clouds obstruct the Sun. (**BR**) Fossilized skeleton of a dinosaur.*

Screaming Cold Wind

Another natural phenomena that God chose to use is the cold screaming wind, which He used for the destruction of the people of 'Ād (41:16). The word *ṣarṣar*, which is used to describe the wind, means "cold screaming."

The Qur'anic Description

The wind that was sent to 'Ād has several characteristics. At first, they were pleased to see a cloud approaching their valley. They thought it was a rain cloud (46:24), but instead, it was a devastating wind (51:41) — cold with a tremendous screaming sound (41:16 and 54:19), exceedingly violent (69:6), and containing lightening bolts (41:13). God concentrated it on them for seven nights and eight days in succession (69:7). The wind plucked out the people like palm trees pulled from their roots and left cored out clean (54:20). They were laid prostrate in its path like roots of hollow palm trees tumbled down (69:7). There were no survivors (69:8). Everything that the wind came up against was reduced to ruin (51:42). It destroyed everything and left only the ruins of their houses (46:25).

Then, when they saw the (penalty in the shape of) a cloud traversing the sky, coming to meet their valleys, they said, "This cloud will give us rain! "Nay, it is the (calamity) you were asking to be hastened! — a wind wherein is a grievous penalty! "Everything will it destroy by the command of its Lord!" Then by the morning they — nothing was to be seen but (the ruins of) their houses! Thus do We recompense those given to sin! (46:24–25)

فَلَمَّا رَأَوْهُ عَارِضًا مُسْتَقْبِلَ أَوْدِيَتِهِمْ قَالُوا هَـذَا عَـارِضٌ مُمْطِرُنَا بَلْ هُوَ مَا اسْتَعْجَلْتُمْ بِـهِ رِيـحٌ فِيـهَا عَـذَابٌ أَلِيمٌ(٢٤)تُدَمِّرُ كُلَّ شَيْءٍ بِأَمْرِ رَبِّهَا فَأَصْبَحُوا لاَ يُـرَى إِلاَّ مَسَاكِنُهُمْ كَذَلِكَ نَجْزِي الْقَوْمَ الْمُجْرِمِينَ (٢٥)

Surat al-Aḥqāf

And in the 'Ād (people) (was another sign): Behold, We sent against them the devastating wind: It left nothing whatever that it came up against. But reduced it to ruin and rottenness. (51:41–42)

وَفِي عَادٍ إِذْ أَرْسَلْنَا عَلَيْهِمُ الرِّيحَ الْعَقِيمَ(٤١)مَا تَـذَرُ مِـنْ شَيْءٍ أَتَتْ عَلَيْهِ إِلاَّ جَعَلَتْهُ كَالرَّمِيمِ (٤٢)

Surat al-Dhāriyāt

So We sent against them a furious wind through days of disaster, that We might give them a taste of a penalty of humiliation in this life; but the penalty of a Hereafter will be more humiliating still: and they will find no help. (41:16)

فَأَرْسَلْنَا عَلَيْهِمْ رِيحًا صَرْصَرًا فِي أَيَّامٍ نَحِسَاتٍ لِنُذِيقَهُمْ عَذَابَ الْخِزْيِ فِي الْحَيَاةِ الدُّنْيَا وَلَعَذَابُ الْآخِرَةِ أَخْزَى وَهُمْ لَا يُنصَرُونَ (١٦)

Surat al-Fuṣṣilat

For We sent against them a furious wind, on a day of violent disaster, plucking out men as if they were roots of palmtrees torn up (from the ground). (54:19–20)

إِنَّا أَرْسَلْنَا عَلَيْهِمْ رِيحًا صَرْصَرًا فِي يَوْمِ نَحْسٍ مُسْتَمِرٍّ (١٩) تَنزِعُ النَّاسَ كَأَنَّهُمْ أَعْجَازُ نَخْلٍ مُنْقَعِرٍ (٢٠)

Surat al-Qamar

And the 'Ād — they were destoyed by a furious wind, exceedingly violent; He made it rage against them seven nights and eight days in succession: so that you could see the (whole) people lying prostrate in its (path), as if they had been roots of hollow palm trees tumbled down! And then do you see any of them left surviving? (69:6–8)

وَأَمَّا عَادٌ فَأُهْلِكُوا بِرِيحٍ صَرْصَرٍ عَاتِيَةٍ (٦) سَخَّرَهَا عَلَيْهِمْ سَبْعَ لَيَالٍ وَثَمَانِيَةَ أَيَّامٍ حُسُومًا فَتَرَى الْقَوْمَ فِيهَا صَرْعَى كَأَنَّهُمْ أَعْجَازُ نَخْلٍ خَاوِيَةٍ (٧) فَهَلْ تَرَى لَهُم مِّن بَاقِيَةٍ (٨)

Surat al-Ḥāqqah

But if they turn away, say: "I have warned you of a stunning punishment (as of thunder and lightning) like that which (overtook) the 'Ād and the Thamūd! (41:13)

فَإِنْ أَعْرَضُوا فَقُلْ أَنذَرْتُكُمْ صَاعِقَةً مِثْلَ صَاعِقَةِ عَادٍ وَثَمُودَ (١٣)

Surat al-Fuṣṣilat

The Scientific Explanation

None of the violent winds known to man today would fit exactly the description of this wind. Hurricanes and typhoons are tropical cyclones (Figure 12). They have maximum sustained winds of about 120 km/h. Atlantic and Eastern Pacific storms are called hurricanes. The word comes from the West Indian word "huracan" and means "big wind." Western Pacific storms are called typhoons, from the Chinese *taifun* meaning "great wind."

Science Miracles: No Stick or Snakes

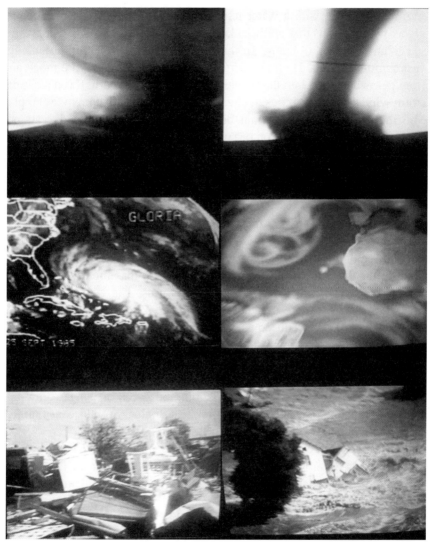

Figure 12: (TL & TR) Tornadoes. *(MR & ML)* Hurricane Gloria initiated in the ocean off the West coast of Africa and spreading towards the United States. *(BL)* Wind destruction. *(BR)* Flood destruction.

The primary energy source for tropical cyclones is the latent heat resulting from water vapor condensation. Only extremely moist air can supply the energy necessary to spawn and maintain tropical storms, and only very warm air contains enough moisture. Tropical cyclones, therefore, only form over oceans with a water temperature of at least 27 degrees centigrade, which means that northern latitudes are normally spared. The warm sea creates a funnel of humid air that rises to perhaps 40,000 ft, producing vast cumulus clouds. High air currents are distributed, and more moist air from below is drawn into the funnel and a hurricane is born. It is a mass of storm winds up to 400 miles in diameter, swirling round at up to 200 mph. The rapidly whirling tangential circulation of hurricane winds can be explained by the conservation of angular momentum. This is similar to an ice skater who spins faster as he/she brings his/her arms down.

The air rotates faster as it is pulled in toward the center of the storm by the low pressure. Hurricanes intensify when they pass over warmer water and weaken when they pass over colder water. The output of power in a single minute could supply to the United States electricity for 50 years. A mature hurricane storm is characterized by an eye at its center (a cloud-free circular region of relatively light wind with a diameter of 10-100 km). The wind in the eye wall rotates counter clockwise at maximum velocity that may exceed 300 km/h in the most severe storms. The wind, waves, and tides generated by it are the cause of damage. The hurricane that struck Bangladesh in 1970 produced a tidal wave that killed at least 500,000 people; and in Galveston, Texas, the hurricane storm tides swept 6,000 people to their death in 1900.

On land, humidity, which is the main power source, is instantly cut off. Forests and mountains slow it by friction, and it can no longer draw up water vapour to power its condensation heat engine. So hurricanes rarely last more than a day on land.

Historically, 'Ād was fourth in generation from Noah, having been a son of 'Aus, the son of Aram, the son of Sam, the son of Noah. They occupied a large tract of land in southern Arabia, extending from Umman at the mouth of the Persian Gulf to Hadhramaut and Yemen at the southern end of the Red Sea. Their location is quite suitable for a visitation of a tropical cyclone, however, being on land its continuation for eight days and seven nights makes it more violent than the hurricane winds presently known to us.

Earthquakes

Earthquakes are known to have occurred on Earth since its beginning. Earthquakes were put on record in the Qur'an to show their importance as a fact of life. It shows their danger. Like wind, it is another power that God

Science Miracles: No Stick or Snakes

reserves the right to use anytime He so wishes. Most of the world's population live in big cities, the majority of which are built on areas known to have seismic activity (Figure 13). Great losses in human life have been recorded in these places. An earthquake, before dawn July 28 1976, in Tanshan, north China, killed approximately 250,000 people.

The Qur'anic Description

The Qur'an describes how "the earthquake" started before the morning accompanied by a blast (11:67; 54:31). It was also accompanied by lightening bolts (41:13; 51:44). The people of Thamūd were found in the morning prostrate in their homes (7:78) like dry stubble used by herdsmen to make pens for enclosure of cattle (54:31).

The (mighty) blast overtook the wrongdoers, and they lay prostrate in their homes before the morning. (11:67)

وَأَخَذَ الَّذِينَ ظَلَمُوا الصَّيْحَةُ فَأَصْبَحُوا فِي دِيَارِهِمْ جَــاثِمِينَ (٦٧)

Surah Hūd

For We sent against them a single mighty blast, and they became like the dry stubble used by one who pens cattle. (54:31)

إِنَّا أَرْسَلْنَا عَلَيْهِمْ صَيْحَةً وَاحِدَةً فَكَانُوا كَهَشِيمِ الْمُحْتَظِــرِ (٣١)

Surat al-Qamar

But if they turn away, say: "I have warned you of a stunning punishment (as of thunder and lightning) like that which (overtook) the 'Ād and the Thamūd!" (41:13)

فَإِنْ أَعْرَضُوا فَقُلْ أَنذَرْتُكُمْ صَاعِقَةً مِثْلَ صَاعِقَةِ عَادٍ وَثَمُــودَ (١٣)

Surat al-Fuṣṣilat

But they insolently defied the command of their Lord: so the stunning noise (of an earthquake) seized them, even while they were looking on. (51:44)

فَعَتَوْا عَنْ أَمْرِ رَبِّهِمْ فَأَخَذَتْهُمُ الصَّاعِقَةُ وَهُمْ يَنظُرُونَ (٤٤)

Surat al-Dhāriyāt

So the earthquake took them unawares, and they lay prostrate in their homes in the morning! (7:78)

فَأَخَذَتْهُمُ الرَّجْفَةُ فَأَصْبَحُوا فِي دَارِهِمْ جَاثِمِينَ (٧٨)

Surat al-A'rāf

Scientific Explanation

An earthquake is a naturally induced shaking of the ground, caused by the fracture and sliding of rock. Six thousand are detected annually in the world. Five thousand five hundred are small or occur in unpopulated areas. Four hundred and fifty can be felt with no damage and 35 cause minor damage. The remaining 15 cause great losses of life and property. The average death toll in the 20th century has been 20,000 people annually. This is due to the collapse of houses, bridges, and other structures.

The size of an earthquake is determined by the dimensions of the rupturing fracture or fault, and by the total amount of slip or displacement. The larger the fault or slip the greater the energy released during the earthquake. The energy produces shaking and seismic waves — P and S (pressure and shear) — which radiate throughout the earth. Shaking at the site of the earthquake lasts only a few seconds while the fault ruptures. It may last a few minutes. The generated seismic waves continue to propagate after the movement on the fault has stopped, spanning the globe in 20 minutes. Only in the immediate vicinity of the fault, at the earthquake's epicenter, are these vibrations powerful enough to cause damage. Seismic waves measured by seismometers are detected at great distances from the epicenters. Earthquakes are ranked by the modified Mercalli scale according to intensity, from I (barely felt) to XII (total destruction). The Richter scale came into use in 1935. Destruction of intensity level IX or X on the Mercelli scale registers more than 6.5 on the Richter scale.

Earthquakes are common rather than extraordinary events, reflecting the slow and continuous motion of material within the Earth. They occur at the boundaries of the lithosphere plates (Figure 13). Half of the earthquakes in the world, and the strongest, occur at the borders of the Pacific plate, which stretches 40,000 km around the circumference of the Pacific Ocean. It is a heavily populated area which includes Japan and the West Coast of America.

Since plate boundaries are also the site of most of the world's volcanoes, earthquakes and volcanoes tend to occur in the same area, for example, on the Pacific "Ring of Fire."

At present, there are maps for areas of high potential seismic activity. Their repeated occurrence are recorded regularly and published in local newspapers, e.g., in San Francisco. In such places, houses are built on floats of concrete and rubber and their frames are made from wood of great elasticity. Houses are built of minimum number of floors; however, with the progress of science, high rise buildings can be seen in big cities like Los

Science Miracles: No Stick or Snakes

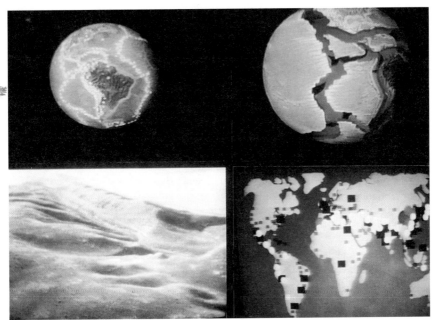

Figure 13: (TL) Areas of earthquakes in the world that correspond to *(TR)* the borders of the plates of the lithosphere (theory of plate tectonics). *(TR)* Please note the "Ring of Fire" where 50 percent of the earthquakes occur around the Pacific plate. *(BL)* San Andreas fault, California. *(BR)* Earthquakes in big cities (Yellow = 10,000 died; Black = 50,000 died; Red = >300,000 died)

Angeles and San Francisco. Warning systems are becoming more sophisticated. Pre-warnings are given not only by the press, but also by radio and television.

Historically, the people of Thamūd occupied the northwest area of the Arabian peninsular between Medina and Syria. This is an area that has always been the site of seismic activity. Thamūd was the son of 'Abir (which means smells good) and a brother of Aram, the son of Sam, who was the son of Noah.

Day of the Overshadowing Cloud

Overshadowing clouds are known to occur as a result of erupting volcanoes. The Companions of the Wood (15:78) were referred to as the Madyan people (7:85). They were probably living in the Sinai area which has a history of volcanic and earthquake activity.

The Qur'anic Description

Before the morning, a blast took place (11:94) accompanied by an earthquake (29:37) then their city was overshadowed with a cloud of a day of great torture (26:189). They were left prostrate in their homes (29:37; 7:91; 11:94).

When Our decree issued, We saved Shu'ayb and those who believed with him, by (special) mercy from Ourselves: But the (mighty) blast did seize the wrongdoers, and they lay prostrate in their homes by the morning. (11:94)	وَلَمَّا جَاءَ أَمْرُنَا نَجَّيْنَا شُعَيْبًا وَالَّذِينَ آمَنُوا مَعَهُ بِرَحْمَةٍ مِنَّا وَأَخَذَتِ الَّذِينَ ظَلَمُوا الصَّيْحَةُ فَأَصْبَحُوا فِي دِيَارِهِمْ جَاثِمِينَ (٩٤) Surah Hūd
But they rejected him: then the mighty blast seized them, and they lay prostrate in their homes by the morning. (29:37)	فَكَذَّبُوهُ فَأَخَذَتْهُمُ الرَّجْفَةُ فَأَصْبَحُوا فِي دَارِهِمْ جَاثِمِينَ (٣٧) Surat al-'Ankabūt
But they rejected him. Then the punishment of a day overshadowing gloom seized them, and that was the penalty of a great day. (26:189)	فَكَذَّبُوهُ فَأَخَذَهُمْ عَذَابُ يَوْمِ الظُّلَّةِ إِنَّهُ كَانَ عَذَابَ يَوْمٍ عَظِيمٍ (١٨٩) Surat al-Shu'arā'
But the earthquake took them unawares, and they lay prostrate in their homes before the morning! (7:91)	فَأَخَذَتْهُمُ الرَّجْفَةُ فَأَصْبَحُوا فِي دَارِهِمْ جَاثِمِينَ (٩١) Surat al-A'rāf

Scientific Explanation

It is easy to understand the occurrence of earthquakes with a fatal outcome. Blasts can occur as a result of earth movement. Those who hear it in active seismic areas, like New Zealand, tell me it sends a shiver up their spine. However, the great torture overshadowing them may be metaphorical. If it actually took place, it could be explained by volcanic eruption with a mushroom cloud, dropping on them its contents, which could be brimstone or any other destructive material. The overshadowing of their city might have resulted from outer space. The historical fires of Peshtigo in 1871 which killed 1,200 people were thought to be due to meteorites. They burned in the Earth's atmosphere setting fire to the woods and houses, covering the city with smoke leading to the total destruction of the city (Figure 14).

Science Miracles: No Stick or Snakes

Figure 14: (TL) Meteorite entering Earth's atmosphere and burning up. (TR) Plaque commerating Peshtigo Fire Cemetery. (ML,MR) Fires (BR) Smoke. (BL) Simulated fire in a large city.

Splitting of the Water

The only natural phenomenon that possesses the power and precision of a surgical blade is the tornado. It starts as a funnel cloud that rotates gently then twists violently within minutes. The funnel then becomes more organized and descends further from the cloud, sometimes touching the ground. Ground contact is often of an intermittent nature. In its center, air surges upwards at 100 to 200 mph, which creates enough suction power to drag objects, as big as cars and wooden houses, into the whirling spout, where they spin to the top and fly out again (Figure 12). Waterspouts are funnel clouds extending from the bases of cumulonimbus clouds to the sea surface of equatorial oceans and inland seas and like tornatoes they can cause serious damage. They develop by a similar process as tornadoes.

The Qur'anic Description

God asked the Prophet Moses to strike the sea with his rod. It divided and each separate part became like a huge mountain (26:64). Pharaoh and his soldiers chased Moses and his people in insolence and spite until water caught up with Pharaoh and he was drowning and he said, "I believe there is only one god whom the Israelites believe in, and I am a Muslim" (10:90; 2:50). The traditional explanation is that the sea was split into twelve passages where the sides of each were like a big mountain. None of those who passed through had a wet saddle or wet clothes. The twelve passages were presumably created for the twelve tribes of Israel. This obviously would shorten the time of crossing.

Then We told Moses by inspiration: "Strike the sea with your rod." So it divided, and each separate part became like the huge, firm mass of a mountain. (26:63)

فَأَوْحَيْنَا إِلَى مُوسَى أَنِ اضْرِبْ بِعَصَاكَ الْبَحْرَ فَانْفَلَقَ فَكَانَ كُلُّ فِرْقٍ كَالطَّوْدِ الْعَظِيمِ (٦٣)

Surat al-Shu'arā'

We took the Children of Israel across the sea: Pharaoh and his hosts followed them in insolence and spite. At length, when overwhelmed with the flood, he said: "I believe that there is no god except Him whom the Children of Israel believe in: I am of those who submit (to God in Islam)." (10:90)

وَجَاوَزْنَا بِبَنِي إِسْرَائِيلَ الْبَحْرَ فَأَتْبَعَهُمْ فِرْعَوْنُ وَجُنُودُهُ بَغْيًا وَعَدْوًا حَتَّى إِذَا أَدْرَكَهُ الْغَرَقُ قَالَ آمَنْتُ أَنَّهُ لاَ إِلَـهَ إِلاَّ الَّذِي آمَنَتْ بِهِ بَنُو إِسْرَائِيلَ وَأَنَا مِنَ الْمُسْلِمِينَ (٩٠)

Surah Yūnus

And remember We divided the sea for you and saved you and drowned Pharaoh's people within your very sight. (2:50)

وَإِذْ فَرَقْنَا بِكُمُ الْبَحْرَ فَأَنْجَيْنَاكُمْ وَأَغْرَقْنَا آلَ فِرْعَوْنَ وَأَنْتُمْ تَنْظُرُونَ (٥٠)

Surat al-Baraqah

Scientific Explanation

Tornadoes derive their name from the Spanish *tronada* meaning "thunderstorm." They are also called "twisters" or "cyclones." They are rapidly rotating columns of air from cumulonimbus clouds, generally observed as tube or funnel-shaped clouds. At ground level, they usually leave a path of destruction only about 50 m (170 ft) wide and travel at an average speed of 8-24 km/hr (5-15 miles/hr) but speeds greater than 30 m/sec (100 ft/sec) have been recorded. The tornado center is an area of exceedingly low pressure that can cause buildings to explode. Wind speeds of approximately 800 km/h (500 m/h) have been inferred from the resultant damage. They are classified on the Fujita-Pearson scale.

The United States experiences about 1,000 tornadoes per year. The strongest on record was in March 1925 and the most outstanding outbreak was 148 tornadoes which occurred between 1:10 pm on April 3rd 1974 and 5:20 am on April 4th 1974 in the Midwest United States. Tornadoes are known to happen in multiples as a family.

When Moses reached the Red Sea, God asked him to strike the sea with his rod. Presumably, the Prophet Moses was heading toward the Sinai and the Gulf of Suez. The Red Sea is a tidal sea. If a family of strong tornadoes, arising from cumulonimbus clouds, developed right above them and was moving with the prevailing north easterly wind, they could possibly create the passages. The funnels of the tornadoes have a suction power of up to 200 m/h which could easily make several clear passages in the water for the twelve tribes of Israel. Each tornado with its suction power will create, directly under it, a mountain of water as mentioned in the Qur'anic description. The continuous suction will create a passage on either side of the water mountain. If another two tornadoes suitably distanced on either side of the first tornado, each will create another mountain of water and help in clearing a passage on either side of the first one. In other words every three adjacent tornadoes, suitably spaced, would make two absolutely clear passages of land for Moses's followers to use. The twelve passages would, theoretically, be made by a family of thirteen tornadoes.

As the tornadoes leave the water and travel overland, they become slower, lose part of their power dropping the mountainous water. These would come down on Pharaoh's soldiers who were chasing behind. Moses's followers would have crossed safely in good time.

Many of the Prophet Moses's miracles can be explained on scientific grounds. Insects like locusts and lice have been known to invade a land. Epidemics in people and animals are well documented. The presence of bloody water in the Nile could be due to gross contamination with iron oxide ore, which is abundant in the area of Aswan. Iron mines are now used in this area for the manufacture of steel. The area is the site of repeated floods that would drain the iron ore into the Nile making it look like blood. Nile water carries clay from Eritrea and contains biological substances that, in the presence of iron, would taste like blood.

Man As Successor of God on Earth

God has chosen man to be His successor on earth (2:30; 27:62). This puts a great responsibility on his shoulders. His tasks are numerous. He has to develop the Earth to make it an ideal place to live in. Scientific references in the Qur'an are there for us to find out their significance. We must develop our scientific knowledge and technology to understand the manner in which the different objects in the universe move or function. So far, we have been successful to imitate many things: light, flying, moving underwater, etc. We have a lot to learn and to uncover. Perhaps the word uncover is more correct than discover. Everything is already there; all that we have to do is find out how things function in nature.

Behold, your Lord said to the angels: "I will create a vicegerent on earth." They said: "Will You place therein one who will make mischief therein and shed blood? While we do celebrate Your praises and glorify Your holy (name)?" He said: "I know what you know not." (2:30)

وَإِذْ قَالَ رَبُّكَ لِلْمَلَائِكَةِ إِنِّي جَاعِلٌ فِي الْأَرْضِ خَلِيفَةً قَالُوا أَتَجْعَلُ فِيهَا مَنْ يُفْسِدُ فِيهَا وَيَسْفِكُ الدِّمَاءَ وَنَحْنُ نُسَبِّحُ بِحَمْدِكَ وَنُقَدِّسُ لَكَ قَالَ إِنِّي أَعْلَمُ مَا لَا تَعْلَمُونَ (٣٠)

Surat al-Baraqah

Or, who listens to the (soul) distressed when it calls on Him, and who relieves its suffering, and makes you (mankind) inheritors of the earth? (Can there be another) god besides God? Little it is that you heed! (27:62)

أَمَّنْ يُجِيبُ الْمُضْطَرَّ إِذَا دَعَاهُ وَيَكْشِفُ السُّوءَ وَيَجْعَلُكُمْ خُلَفَاءَ الْأَرْضِ أَإِلَهٌ مَعَ اللَّهِ قَلِيلًا مَا تَذَكَّرُونَ (٦٢)

Surat al-Naml

The biggest challenge is to protect our Earth from the natural phenomena that originate on our little planet. We are still helpless when it comes to big floods, major earthquakes, and unexpected tsunamis. We would be even more helpless if we were to be bombarded by meteorites or asteroids. Earth-based observatory stations and satellites have reported about 250 meteorites over the last ten years (Figure 9) which have detonated at the Earth's atmosphere. One of them is reported to have produced the explosive power equal to four or five times the atomic bomb that fell on Hiroshima. If a meteorite or asteroid of 5 miles in diameter size fell on the United States, it would cause its complete destruction (Figure 9). Asteroids capable of destroying a quarter of the world's population collide with Earth twice every million years. Smaller objects capable of destroying the population of a major city could hit once every two or three centuries. Impacts, therefore, put the whole planet at risk; however, this is the only natural disaster that we may be able to do something about.

A scientific meeting held under the title of "The End of the World" was attended by representatives of many countries. With an objective to safeguard the Earth against invasion by meteorites, asteroids or comets, it was suggested that the strongest long-range rocket in the world, the Russian Enagia, could be used to carry an American warhead to divert an asteroid or meteorite approaching Earth. Changing slightly its orbit away from the Earth would save us from a definite disaster. The destruction of the meteorite or asteroid, however, might prove catastrophic as one of its fragments may collide with the Earth (Figure 15).

Historical Universe Natural Phenomena, Science Miracles

The Qur'anic description of the universe's natural phenomena is further evidence of scientific miracles in the Qur'an. The description is accurate and forewarns us of possible dangers. The Prophet Muhammad could neither read nor write. The information could only have been produced from a scientific encyclopedia of the twentieth century.

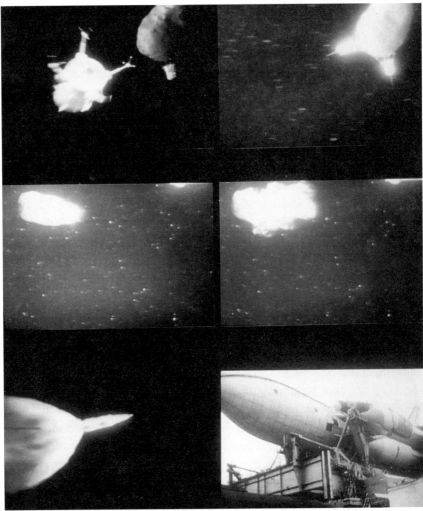

Figure 15: *(**BR**) The Russian Enagia, the strongest long range rocket in the world, can carry an American nuclear head (**BL**) to be detonated on the surface of an asteroid thus avoiding its collision with the Earth (**TR & TL**) or destroynig it (**ML & MR**).*

CHAPTER ELEVEN

INITIATION OF SIGNATURE

So far, the symbolic letters have led to a particular path of knowledge in the Qur'an. They seem to build up to something outstanding for mankind. At first, we saw how the *nūn* spreads to make a wide base and a foundation for the miraculous language of the Qur'an. We shall see later, that it also stands as the foundation and the final light point of God's signature of His name, the Merciful (*Al-Rahmān* الرحمن). We have seen how *qāf* (ق) builds our scientific background of the universe and man. How man's existence does not go unrecorded; he has to answer for his actions. His record is kept by a concealed, sophisticated, and fair system, probably in his brain. God has shown us, as promised, how He chooses the prophets who carry His message to man. His specified system of guidance requires patience. He asked the Prophet Muhammad to be patient and set Himself as an example.

People should live in peace and believe in life after death. They have to give an account of what they have done during their lifetime. They should realize that they have already testified that God exists, long before they came to life, so it is no good trying to deny it in the current life. There is no need to doubt His power of creation. They can see many of His signs in the universe and also in themselves. He can regenerate senile organs and can begin reproduction without sexual union, parthenogenesis. He warns us that the Earth is not on its own in the universe. There are natural phenomena, on Earth and in the universe, that are a potential danger to our existence. Man, being God's successor, must learn how to protect himself by hard work, studying His signs, and research. God has helped man a great deal to survive such phenomena. He described them accurately so that we would fear Him, be aware of their danger, and be prepared.

God initiates the signature of His name, The Merciful (*Al-Rahman* الرحمن), by inscribing the three letters *alif, lām,* and *rā* (الر). He placed them at the beginning of five surahs (10, 11, 12, 14, 15 or 51, 52, 53, 54, 72 cronologically). No one needs to wonder where the rest of the name is. The middle letters, *ḥā* and *mīm* (حم), are discussed in the next chapter. The name ends with the letter *nūn*. We have seen in Chapter Two that more than half of the verses of the Qur'an end with the letter *nūn*.

The symbolic letters *Alif Lām Rā* refer us to *surah* 55, The Merciful (*Al-Rahman* الرحمن). This *surah* indicates that God is merciful to man as He created him, taught him the Qur'an, and gave him the power of intelligent speech (55:1–4). God begins *surah* 13 (96 cronologically) with the letters *alif lām mīm rā* (المر), which are constituents of the same name. In this *surah*, God declares His name as The Merciful (13:30).

Thus have We sent you among a people before whom (long since) have (other) peoples (gone and) passed away; in order that you mightest rehearse unto them what We send down unto thee by inspiration; yet do they reject (Him), the Most Gracious! Say: "He is my Lord! There is no god but He! On Him is my trust, and to Him do I turn!" (13:30)

كَذَلِكَ أَرْسَلْنَاكَ فِي أُمَّةٍ قَدْ خَلَتْ مِنْ قَبْلِهَا أُمَمٌ لِتَتْلُوَ عَلَيْهِمُ الَّذِي أَوْحَيْنَا إِلَيْكَ وَهُمْ يَكْفُرُونَ بِالرَّحْمَنِ قُلْ هُوَ رَبِّي لاَ إِلَهَ إِلاَّ هُوَ عَلَيْهِ تَوَكَّلْتُ وَإِلَيْهِ مَتَابِ (٣٠)

Surat al-Ra'd

(God) Most Gracious! It is He who has taught the Qur'an. He has created man: He has taught him speech (and intelligence). (55:1–4)

الرَّحْمَنُ(١)عَلَّمَ الْقُرْآنَ(٢)خَلَقَ الإِنْسَانَ(٣)عَلَّمَهُ الْبَيَـــانَ(٤)

Surat al-Raḥmān

Signs Are Miracles and/or Verses

God describes the miracles that He gave to the prophets as *ayāt*. In this context, *ayāt* means "miraculous signs." For instance, God gave the Prophet Moses the *ayah* (singular of *ayāt*) of his stick changing into a snake. On the other hand, each verse in the Qur'an is also called an *ayah*. These verses represent God's revelation, which have been reduced to writing or inscriptions. They include heavenly advice and knowledge that is probably recorded in light, in *umm al-kitāb* (the Mother Book) kept in heaven. It had to be inscribed to be sent to Earth as holy books for man to read.

Science Miracles: No Stick or Snakes 93

The contents of the verses are so valued in knowledge that they may be considered as new information, more or less equal to miracles for mankind. The first verse of *surah*s 10 through 15 give this explanation. The merciful God has sent His manifest book with signs to bring mankind out of darkness into light. But this can only be done by following His path and studying His signs in detail. The signs appear in the Qur'an as scientific allusions. Those recorded in the *surah*s under study will be given later.

Selective Saving from Mass Destruction

Creation of Earth went through many difficult stages. Man was put on Earth without canine teeth, claws, or physical power that would qualify him to be master of the Earth; however, God gave him a unique brain and made him His successor on Earth (2:30). Many obstacles and natural disasters befell him. He could not have survived them without the kind help of the Merciful. God allows people to taste His mercy (10:21; 11:9); He did not have to guide us. He could have replaced us (47:38 & 5:54). Being the Merciful, He sent the prophets — Moses (11:17), Noah (11:28), Salah (11:63), Shu'ayb (11:90) — with the holy books for guidance and knowledge. He saved the people who followed them from His punishment by natural phenomena, again through His mercy — Moses (10:86), Hūd (11:58), Thamūd (11:66), Shu'ayb (11:94). The human soul incites one to evil and is only tamed by the grace and mercy of God (12:53). He is forgiving and visits His mercy on whom He wishes (12:56,64,92,98; 14:36; 15:49,56).

Scientific Allusions as Science Miracles

Scientific allusions are given in the Qur'an rather than miracles. God was well aware that He would open His gates of knowledge for the progress of science to the nation of Muhammad. A stick that transforms into a snake or raising one or two people from the dead would not continue to impress people after the millennia had passed. Scientific allusions were given instead. When their significance is found, God's presence will be recognized (27:93).

Scientific Allusions Related to the Universe

God originated and created the heavens and the earth in six days of the universal clock, referring to the fourth dimension (10:3; 14:10), and His throne was on water (11:7). It has also been recently shown and published in the *Guardian,* Wednesday April 8 1998, that the European scientists using an ultracold orbiting telescope have discovered unimaginable volumes of water in the space between the stars. Water vapor was found in the freezing atmosphere of Jupiter, Uranus, Neptune, and Saturn. They also discovered vast oceans of water vapor the mass of which in the Milky Way galaxy alone would equal that of tens of thousands of suns. They have even identified a cloud of water, less than a light year across, in the constellation Orion. Early in the creation, particles of hydrogen were created, which is the most abundant element in the universe. Once the particles of hydrogen are present, other elements can be made by nucleosynthesis in the stars where the temperature is so high as to allow them to act as furnaces where this process can be completed. Oxygen is made in huge quantities by the stars. So water (composed of hydrogen and oxygen) should not be a surprise.

Before the Day of Judgment, the heavens and earth will change (14:48) but after this event, they will not change (11:107–108). If God were to open a door into heaven for people, they would think they were bewitched (15:14–15). He raised the heavens without any seen pillars and forced the Sun and the Moon to run to a known life span (13:2). Everything He created is balanced (15:19). He has the stores of everything and He causes it to descend in known quantities (15:21). He created the Sun as a source of light and the Moon to shine. The Moon is made to traverse houses (days or stages) to teach man arithmetic and the calendar. Arithmetic began with celestial calculations (10:5). He created constellations for guidance at night (15:16). Among them are manifest flames that can pursue like rockets (15:18). Night and day alternate. The alternating of night and day and what is created in the heavens and earth are signs for man (10:6)

Scientific Allusions Related to the Earth

He extended the Earth and threw in it *rawasi* or lithosphere (15:16,19; 13:3). He reduces it from its extremities (13:41) at the trenches in the Pacific. The mountains act as pegs (78:7) which He has berthed (*arsa*) (79:32). He differentiated the Earth into layers (65:12). (This alludes to the recently discovered theory of plate tectonics. There are pieces in the Earth neighboring each other that could be plates or something yet to be discovered.)

All plants are irrigated by the same water but they all vary in taste (13:5), and He made pairs of all fruits (13:3). He uses the wind for the transfer of pollens; He drops rain for drinking; and He has all the water stores (15:22).

Science Allusions Related to Man

He created man from a clay of molded mud and blew into him from His spirit (15:26,29). He created jinn from smokeless fire (15:27). He chose man to be His successor on earth (10:14).

God recognizes human instincts like jealousy (12:8), murder (12:9), cruelty (12:10), lying (12:16–18) and sexual temptation (12:24,31–33; 15:39). Nobody can remove evil from man or withhold benefit going to him except God (10:107).

Dreams of man are in code. Their interpretation can only be made by those with vision (12:36,41,43,46–49,100). Hysterical blindness is described (12:93). God knows what every female bears, when a womb is full or empty and everything with Him has its measure (13:8). God mentions a system of protection for man which may be his immune system (13:11).

The Universe's Natural Phenomena

- *Far al Tanur* or Noah's flood. The waves were like mountains and the sea level rose to a dangerous level (11:40–43).
- The Cry or Blast, destroyed the people of Thamūd (11:67; 15:83).
- The Cry or Blast, destroyed the people of Lot (11:82,83) and rained on them stones (15:73,74).
- The Cry or Blast, destroyed the people of Shu'ayb (11:94).

Heavenly Books

The Qur'an is an Arabic language miracle. Nobody can imitate it (11:13–14; 12:2); and God will protect it (15:9).

God keeps in a book everything in the heavens and earth of an atom's weight or less (10:61). Every life has a book (13:38); God removes or establishes what He wants, and He has *umm al-kitāb* (13:39).

God's Power

God brings life out of death and brings death out of life (10:31,55–56; 15:23–24). He creates life and recreates it (10:34). He regenerates senile organs (14:39). He knows what we think to ourselves and what we declare openly (11:5). Every creature has a sound as it moves on Earth, it is provided for by God, and He knows its lodging place and repository. All are in a manifest book (11:6). His blessings cannot be numbered (14:34).

Everything in the heavens and earth kneels to God, either themselves or their shadows (13:15). To Him belong the unseen and the unknown (11:123). He watches every soul (13:33) and does whatever He wants (14:27). He sends lightning to affect whom He wants (13:13). He knows who comes first and who comes last and He inherits the universe (15:23–24).

CHAPTER TWELVE

THE MERCIFUL (AL-RAḤMĀN) PREVAILS THOUGHOUT THE QUR'AN

In this chapter, we describe how God completes His signature by putting the two letters *ḥā* and *mīm* (حم), representing the middle of His name, The Merciful (*Al-Rahman* الرحمن), at the beginning of seven *surahs* (40–46 or 60–66 cronologically). The figure seven is frequently used by God. He fashioned the heavens into seven layers and from the earth their like (65:12). *Surah* 1 (5 cronologically), which opens the Qur'an and is used by mankind for prayers several times a day, is made of seven verses. In the previous chapter, the beginning of the name — *alif lām rā* (الر) — was given. The *nūn* (ن) is present at the end of 3,123 verses of the Qur'an (50.14 percent of the verses). The heart or the middle of the name, *ḥa mīm* (حم) is now in place and the name, the Merciful (*Al-Rahman* الرحمن) is complete. One can easily notice that *alif lām rā ḥā mīm* (الر حم) are placed in the center of the Qur'an, according to cronology. The *surah* numbers cover 50–54 for *Alif Lām Rā* (الر) and 60–66 for *ḥā mīm* (حم). Their central position in the Qur'an facilitates an easy circular access to the letter *nūn* present at the end of 3,123 verses.

The symbolic letters *ḥā mīm* constitute verse 1 of the previously mentioned seven *surahs*. God's name, The Merciful (*Al-Rahman* الرحمن) or its adjectival form appears in all seven *surahs* (40:7; 41:2,32,50; 42:28,48; 43:17,19,20,32,33,36,45,81; 44:6,42; 45:20,30 and 46:8,12)

The letters *'ayn, sīn,* and *qāf* (عسق) constitute verse 2 of *surah* 42. They are the beginning of several of His names that appear in the same *surah*; for example:
- *'Ayn* (ع) is the first letter of *Al-'Azīz*, The Eminent (42:3,19), *Al-'Alīm,* The All-Knowing (42:12,24,25), *Al-'Aẓīm*, The Tremendous (42:4), *Al-'Afuw,* The Forgiving (42:25), *Al-'Adl*, The Just (42:15), and *Al-'Aliy,* The High (42:51);

- *Sīn* (س) is the first letter of *Al-Samiʻ*, The All-Hearing (42:11);
- *Qāf* (ق) is the first letter of *Al-Qādir*, The All-Powerful (42:9), and *Al-Qawī*, The Strong (42:19).

The Merciful has made the Qur'an a miraculous Arabic book (41:3) in which He reveals His scientific signs to us in order that we recognize Him and believe in His existence.

Scientific Allusions as Signs, Miracles (*Surah*s 40–46)

Spirit from His Command

He inspired Muhammad with a spirit from His command to give him the Qur'an (42:52). He made it light to guide His subjects to the right way (42:52). He casts the spirit from His command upon whom He wishes that He may warn people of the Day of Judgment (40:15). He does not speak to man except by revelation or from behind a barrier (42:51).

Physiology

Some of the physiological signs of fear such as choking with anguish and feeling the heart in the throat with palpitations are described in the Qur'an (40:18).

Scientific Allusions Related to the Universe

God asks us to reflect on the creation of the heavens and earth; on the creatures that make sound as they move on the Earth; on the variation of day and night; on the rain which gives provision and revives the dead land; and on the wind (45:3–5,13). The creation of the heavens and earth was greater than the creation of man (40:57). The heavens and earth were smoke and He separated them (41:11). He differentiated the heavens into seven layers in two days of the universe clock and inspired in each its command; and He adorned the lower heaven with lamps to preserve (41:12). He did not do this in play (44:38), and it did not cause Him any fatigue (46:33). God knows the life span of all His creations (46:2). A day will come when the skies will bring manifest smoke that will cover the people and will cause extreme torture (44:10–11),

Science Allusions Related to Earth

He created the earth in two days of the universe clock (41:9) and created *rawasi* (lithosphere) on top of it and made the food in it in four days of the universe clock (41:10).

Science Allusions Related to Man

He created us from dust then from semen to become an *'alaqah* (blood clot). He images us in the womb in the best possible picture (40:64). When a child is born and is looked after by its mother, it remains with her for 30 months (46:15). God decides who will have a male or female child and who will be barren (42:49). He advises consultation among people for making decisions (42:38). People think they only live and die and are destroyed by time (45:24). We die on a specific day known only to God (40:67).

The Universe's Natural Phenomena

He used lightening bolts to destroy the people of 'Ād and Thamūd (41:17). 'Ād also had a clamorous wind (41:16). They saw a cloud and thought rain was coming but it was wind that caused painful chastisement and destroyed everything except their dwelling places (46:24,25).

He made the seas for ships to sail in with winds from His command (45:12,32–33).

Heavenly Books and God's Power

Every nation has a book (45:28,29). He gives life and causes death and says, "Be," and it is (40:68). He is capable of gathering all creatures that can make sound as they move on the Earth from the beginning to the end of creation (42:29). He revives dead land with rainwater (43:11).

He knows the treachery of the eye, which is an uncomprehended lie detection test, and what we conceal in our minds (40:19). God has messengers with us who write down our secrets and what we conspire (43:80). Those who do not believe in God are assigned a satanic companion (43:36).

God Has No Son or Human Parts

God told Muhammad to say to people, if the All-Merciful had a son I would be the first of the worshippers (43:81). God cannot possibly have a human part as an extension (43:15).

God shows us all of these signs, so how can we possibly deny Him (40:13,81).

CHAPTER THIRTEEN

THE BODY OF KNOWLEDGE AND ADVICE

In this chapter six *surah*s are discussed (2, 3, 29, 30, 31, and 32; or 57, 75, 84, 85, 87 and 89 cronologically). They contain 674 verses, 10.8 percent of the Qur'an. They give advice for mankind, put man on the straight path, and correct all the misconceptions prevalent at that time. The symbolic letters *Alif Lām Mīm* (الم) represent the first verse of these *surah*s.

There are twenty-six names of God which begin with *alif lām mūm*. None of these names appear in the above mentioned *surah*s. There are, however, forty-three names of God which contain *mīm* alone. These names or their adjectival forms are repeated in these *surah*s and they are as follows: the Merciful, *Al-Rahman* (34 times); The All-Knowing, *Al-'Alīm* (21 times); The Wise, *Al-Hakīm* (12 Times); The All-Hearing, *Al-Samiʿ* (9 times); the Praiseworthy, *Al-Hamīd* (3 times); and the Avenger, *Al-Muntaqim* (once). From the overall meanings of the names of God and according to their frequency, emphasis is made and the themes of the *surah*s can be seen at a glance. From His mercy, knowledge, and wisdom He gave us many signs to guide us on the right way. He can hear and see our thoughts and actions and He will punish those who hinder or obstruct the way to Him. The *surah*s relate to many social sciences and contain numerous scientific allusions.

Scientific Allusions

Allusions Related to the Qur'an

God revealed the Qur'an to Muhammad. Because he who could neither read nor write no one could claim that it is of human origin (29:48). In Arabic, no one can write a *surah* that compares to a *surah* of the Qur'an

(2:23,24). The Qur'an contains clear signs for those with knowledge (29:49). As one evidence of its heavenly origin, it predicted correctly when the Romans would be defeated (30:2–6).

Allusions Related to Man

God created the first man out of clay, then from mean water (32:8), shaped him and blew into him from His spirit, and finally, appointed for him hearing, sight, and heart (32:7–9). He images him as He wishes inside the womb (3:5,6). He made man His successor (*khalifa*) on earth (2:30,31). He created Jesus by a process reminiscent of parthenogenesis (3:45,47).

Man is easily influenced by the Devil (2:35,36). He has lusts for women, producing children, gold, silver, pedigree horses, cattle and land produce (3:14). However, God only extends His provision to whom He wills (29:62). He will not allow people to destroy the Earth (2:27).

People always admit that it is God who created the heavens and earth, it is He who drops the water from the sky to bring the dead land or earth to life, and it is He who subjects the Sun and the Moon for man's benefit — yet they deny Him (29:61,63). All the examples sent to man can only be understood by those of knowledge (29:43). He to whom God gave the wisdom has been given a great deal (2:269). They will be generously rewarded (2:245,261).

Allusions Related to Public Health and Diagnosis of Pregnancy

For health reasons, God forbids, except when under duress, the eating of carrion, blood, the flesh of swine, and any animal that is killed in the name of other than God (2:173). We must refrain from intercourse during menses (2:222). Breast-feeding should continue for two years (2:233; 31:14). He gives an accurate method for diagnosis of pregnancy (2:228,234) to allow remarriage of divorcees and widows without affecting paternity.

Allusions Related to the Water Cycle

He sends the winds to stir up clouds and spreads them in heaven as He wishes and scatters them to drop rain on some of His servants who will rejoice (30:48). He drives the water to dry land (32:27). Rocks may fracture or crack and water will come out of them to form rivers (2:74).

Allusions Related to the Universe

He created the heavens and the earth in truth, He knows what is in them (29:44,52), their life span (30:8;31:29), and can recreate them (29:19,20; 30:11).

God created the heavens without visible pillars and made them into seven layers (2:28,29). He threw *rawasi* (lithosphere) on earth "lest the earth cave in from under us." His throne of authority stretches over the universe (2:255).

God enters night into day and day into night. He brings the Sun from the East (2:258). He brings life out of death and death out of life and extracts life out of the dead earth — and that is how we will be resurrected. For whom He will, He provides without limitations (3:27; 30:19).

Allusions Related to the Fourth Dimension

God created the heavens and earth in six days of the universe clock then leveled Himself on His throne of authority (32:4). He directs the affairs from heaven to Earth, then it goes up to Him in one day, whose measure is a thousand years of our counting (32:5).

Allusions Related to the Universe's Natural Phenomena

The Universe's natural phenomena have been used for the punishment of man (29:37,40). Man has a instinctive fear of thunder and lightening (2:19,20). During Noah's 950 years among his people, the flood took place (29:14).

God's Power

He created man, with variable language and color, from dust; and from man He created his wives. Man is bound to his mate with love and compassion. He created the universe and can find anything hidden in it (31:16). Man sleeps during the night and during the day seeks His bounty. God shows us lightening in fear and hope that we may recognize Him. He drops rainwater on dead soil to bring life. He holds the heavens and earth by His command. We now know that objects in the universe are rotating in their orbit governed by the laws of gravity. This is perhaps one of many forces, as yet to be discovered, which govern the universe. He will resurrect us after death (30:50). The universe obeys Him and He can recreate us all. His best creation is the heavens and earth (30:20–27). Everything in the universe worships God in a language that we do not understand. God's words are endless (31:27).

CHAPTER FOURTEEN

THE LIGHT CIRCUIT

Reading about the flow of light in the Qur'an attracts everyone's attention and interest. The verses are numerous and they seem to flow as if in a circuit. It starts with God, as the light of the universe, and finishes with man presenting his light to God on the Day of Judgment. One can understand this metaphorically on the basis that knowledge and education are a form of enlightenment. Taking advantage of the progress of knowledge and science, however, we may appreciate God's light as a different and not metaphoric concept. The human body contains complex electric circuits initiated by chemical stimuli, and man has been given energy to perpetuate his life, which could extend to a thousand years, for example the Prophet Noah (29:14). On average, the longevity of man extends seventy to eighty years, according to the will of God. Perhaps, if we put forward how the human body works, a scientific explanation could be made for the light circuit which has been presented by God to man through the prophets and the Qur'an and then back to be presented to God through worthy people.

Man Is Driven by Energy

Origin of the Energy
God created man from mud and put energy into him. His descendants are born with this energy in mean water (32:8). This energy or spirit must have existed before the creation. God brought us all before Him, including those who have not yet been born, and we testified before Him that He is our Creator (7:172). Then the human race started to appear.

Extra Energy for Extra Duties

To the few chosen men and women whom God requires to perform duties over and above their normal capacity He gives extra energy or *rūḥ* from His command, to enable them to undertake their extra responsibilities (40:15).

> Raised high above ranks (or degrees), (He is) the Lord of the Thone (of authority); by His command doth He send the spirit (of inspiration) to any of His servants He pleases, that it may warn (men) of the day of mutual meeting. (40:15)

رَفِيعُ الدَّرَجَاتِ ذُو الْعَرْشِ يُلْقِي الرُّوحَ مِنْ أَمْرِهِ عَلَى مَنْ يَشَاءُ مِنْ عِبَادِهِ لِيُنْذِرَ يَوْمَ التَّلَاقِ (١٥)

Surah Ghāfir

The Weight of Our *Rūḥ*

Our development in science and technology has allowed us to measure the weight of very small objects. Using theories like quantum mechanics, we can estimate weights of minute objects of matter. However, we stop short of weighing the mass of our spirit, which I suggest could be calculated as the difference in the weight of a human being before and immediately after his death. This immeasurable (to us) difference presumably represents the weight of our spirit (*rūḥ*) which leaves our body and is kept with God. This extremely minute energy will disappear from the earth's mass. It is an amount measurable and recognized by God (50:4).

> We already know how much the earth is reduced when they die with us is a record guarding (the full account). (50:4)

قَدْ عَلِمْنَا مَا تَنْقُصُ الْأَرْضُ مِنْهُمْ وَعِنْدَنَا كِتَابٌ حَفِيظٌ (٤)

Surah Qāf

The Dependance of Man on Electrochemical Reactions

As shown in the previous chapters, each cell in the human being possesses a genetic map made of genes dependant on chromosomes and Deoxyribonucleic Acid (DNA). The latter is formed of complex arrangements of amino acids. Each atom in the body carries an electric charge. All electric charges are carried through the body to pass information, to give or take orders. Nerves send electric impulses at the motor endplate to muscles to give them orders to contract or relax. So when one is frightened and wants to run away from danger, the feet do not let one down, while the eyes and ears see and hear the danger. The brain sends orders to the right muscles of the legs and feet and the person starts to run.

The human body is run by electricity. The biological electric potential created in the human body originates from its semipermeable membrane. It is partially permeable to potassium (K^+) and sodium (Na^+). As a result of the cell's electrical properties, a potential of approximately 0.1 volt is generated across the cell's membrane. Changes in potentials of this type are the origin of signals recorded by the Electrocardiogram (ECG), the Electroencephalogram (EEG), and the Electromyogram (EMG).

One of the best ways of understanding the electrical activity of the brain is to study it during sleep, when the patient is free of outside pressures. Every nerve cell produces a detectable electrical voltage whenever it is active. Sometimes, to measure these voltages, we have to go through the scalp. Usually, it is easier to detect the combined activity of many cells. We can detect the total activity of the brain by measuring it through the thickness of the skull. In this situation, a signal detected is made up of separate signals from millions of nerve cells all doing millions of different things. This is like listening to a hundred people, all talking at the same time in a large hall. You could recognize them if they all started saying the same thing or when they stop talking to listen to one of them. The brain wave activity is called the EEG.

During sleep, the brain goes through a series of well-defined stages of activity. The sleeping body continues living — producing hormones, digesting, maintaining the function of the nerves. The brain wave pattern becomes more regular — ten cycles/second — as the brain cells start to act in synchrony (alpha rhythm), thus indicating that the section of the brain connected with attention is switched off. When we pass from wakefulness to sleep our brain produces bigger and slower brain waves than those it produces when we are awake. This continues until deepest sleep is reached at which time the frequency becomes 1–2 cycles/second. This suggests that the cortex, where we take the measurement, has reduced its activity. This lasts about half an hour or so, then very quickly the brain waves start to become more active, that is, of higher frequency. The brain cells, instead of doing the same thing, begin to split up into small groups and become as active as when we are awake. The heart rate goes up and so does the respiratory rate, the eyes move rapidly and there is increased activity of the sex organs with erection and secretion. This is the rapid eye movement (REM) sleep, which lasts 20-30 minutes. This type of sleep will alternate with deeper sleep during the night. The periods of rapid eye movement seem to coincide with the periods when the sleeper is dreaming (Figure 16).

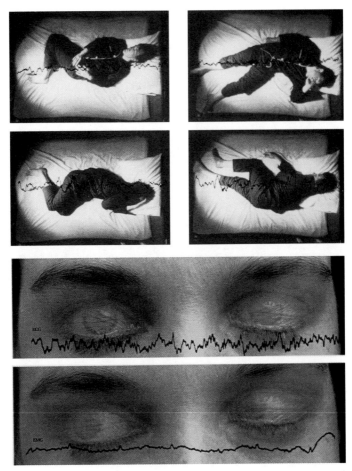

Figure 16: (TL) A few minutes after sleeping, brain waves show irregularities, typical of waking brain. *(TR)* The waves start to become bigger and slower. *(UML&UMR)* The waves become larger and slower. In deep sleep heart rate and breathing are slow and muscles are relaxed. We oscillate between deep and shallower levels of sleep with dreaming sleep between. *(LM&B)* Traces of the ECG and EMG during REM sleep. The EMG (Electromyograph) trace shows an absence of muscle tone, because the brain stem blocks movement messages.

Figure 17 illustrates the Electroencepholagram of a normal person and those with petit mal, major epileptic seizure and cerebral abscess. They are all different in frequency, magnitude, and significance.

From the above examples, one can appreciate that every single thought or movement is recorded by an electric impulse or signal and stored somewhere in the brain.

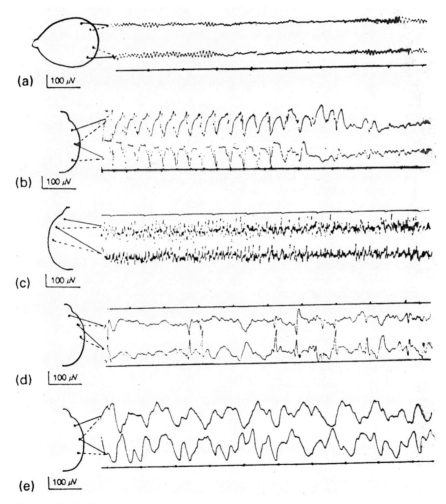

Figure 17: EEG a) normal alpha rhythm, it dissappears in the center when the eyes are open. b) A 3 Hz spike and wave discharge of petit mal epilepsy. c) High frequency discharges (mainly muscle artifact) during a major epileptic seizure. d) A temporal lobe epilepsy. e) A focus of high-amplitude delta activity seen in the right mid-temporal region in a patient suffering from a cerebral abcess.

The Light Circuit Path

The Light Source of the Universe

God is the light of the heavens and earth (24:35). His eternal light is produced without any fire. After the Day of Doom, the Sun and the Moon will disappear and the Earth will be changed into a new one (14:48). This new Earth will shine with the light of God (39:69). This is why one of God's beautiful names is The Light (*Al-Nūr* النور).

God is the Light of the heavens and the earth. The parable of His light is as if there were a niche and within it a lamp; the lamp enclosed in glass; the glass as it were a brilliant star: lit from a blessed tree, an olive, neither of the East nor of the West, whose oil is well-nigh luminous, though fire scarce touched it: light upon light! God doth guide whom He will to His light: God doth set forth parables for men; and God doth know all things. (24:35)

اللَّهُ نُورُ السَّمَوَاتِ وَالْأَرْضِ مَثَلُ نُورِهِ كَمِشْكَاةٍ فِيهَا مِصْبَاحٌ الْمِصْبَاحُ فِي زُجَاجَةٍ الزُّجَاجَةُ كَأَنَّهَا كَوْكَبٌ دُرِّيٌّ يُوقَدُ مِنْ شَجَرَةٍ مُبَارَكَةٍ زَيْتُونَةٍ لاَ شَرْقِيَّةٍ وَلاَ غَرْبِيَّةٍ يَكَادُ زَيْتُهَا يُضِيءُ وَلَوْ لَمْ تَمْسَسْهُ نَارٌ نُورٌ عَلَى نُورٍ يَهْدِي اللَّهُ لِنُورِهِ مَنْ يَشَاءُ وَيَضْرِبُ اللَّهُ الْأَمْثَالَ لِلنَّاسِ وَاللَّهُ بِكُلِّ شَيْءٍ عَلِيمٌ (٣٥)

Surat al-Nūr

And the earth will shine with the glory of its Lord: the record (of deeds) will be placed (open); the prophets and the witnesses will be brought forward; and a just decision pronounced between them; and they will not be wronged (in the least). (39:69)

وَأَشْرَقَتِ الْأَرْضُ بِنُورِ رَبِّهَا وَوُضِعَ الْكِتَابُ وَجِيءَ بِالنَّبِيِّينَ وَالشُّهَدَاءِ وَقُضِيَ بَيْنَهُمْ بِالْحَقِّ وَهُمْ لاَ يُظْلَمُونَ (٦٩)

Surat al-Zumar

The Light Path from Heaven to Earth

God inspired Muhammad with a spirit from His command at a time when he was not aware of the Qur'an. God made the Qur'an a light for guidance to direct mankind to the right way (42:52). Similar light was previously sent to Earth with earlier prophets who delivered God's guidance. Some of them were given holy books (i.e., the messengers). The light of God descended in the holy book of the Prophet Moses and his signs (6:91; 14:5).

Science Miracles: No Stick or Snakes

And thus have We, by Our command, sent inspiration to you: You knew not (before) what was revelation, and what was faith; but We have made the (Qur'an) a light, wherewith We guide such of Our servants as We will; and verily do you guide (men) to the straight way. (42:52)

وَكَذَلِكَ أَوْحَيْنَا إِلَيْكَ رُوحًا مِنْ أَمْرِنَا مَا كُنْتَ تَدْرِي مَا الْكِتَابُ وَلَا الْإِيمَانُ وَلَكِنْ جَعَلْنَاهُ نُورًا نَهْدِي بِهِ مَنْ نَشَاءُ مِنْ عِبَادِنَا وَإِنَّكَ لَتَهْدِي إِلَى صِرَاطٍ مُسْتَقِيمٍ (٥٢)

Surat al-Shūrā

No just estimate of God do they make when they say: "Who then sent down the Book which Moses brought? A light and guidance to man: But you make it into (separate) sheets for show, while you conceal much (of its contents): therein were you taught that which you knew not — neither you nor your fathers." Say: "God (sent it down)"; then leave them to plunge in vain discourse and trifling. (6:91)

وَمَا قَدَرُوا اللَّهَ حَقَّ قَدْرِهِ إِذْ قَالُوا مَا أَنْزَلَ اللَّهُ عَلَى بَشَرٍ مِنْ شَيْءٍ قُلْ مَنْ أَنْزَلَ الْكِتَابَ الَّذِي جَاءَ بِهِ مُوسَى نُورًا وَهُدًى لِلنَّاسِ تَجْعَلُونَهُ قَرَاطِيسَ تُبْدُونَهَا وَتُخْفُونَ كَثِيرًا وَعُلِّمْتُمْ مَا لَمْ تَعْلَمُوا أَنْتُمْ وَلَا آبَاؤُكُمْ قُلِ اللَّهُ ثُمَّ ذَرْهُمْ فِي خَوْضِهِمْ يَلْعَبُونَ (٩١)

Surat al-An'ām

We sent Moses with Our signs (and the command). "Bring out your people from the depths of darkness into light, and teach them to remember the days of God. Verily in this there are signs for such as are firmly patient and constant — grateful and appreciative. (14:5)

وَلَقَدْ أَرْسَلْنَا مُوسَى بِآيَاتِنَا أَنْ أَخْرِجْ قَوْمَكَ مِنَ الظُّلُمَاتِ إِلَى النُّورِ وَذَكِّرْهُمْ بِأَيَّامِ اللَّهِ إِنَّ فِي ذَلِكَ لَآيَاتٍ لِكُلِّ صَبَّارٍ شَكُورٍ (٥)

Surah Ibrāhim

The Origin of the Qur'an

The Qur'an is a revelation from heaven. Some of its verses and signs are fundamental and these are *umm al-kitāb* (the Mother Book). This is a heavenly book, which is probably written in light and contains all the instructions and guidance that has been sent to man. On Earth part of it was transcribed and made into a readable holy book, the Qur'an, that can be studied by mankind (3:7).

The Light in the Qur'an

God describes the Qur'an as High and Wise (43:4). God sent down the Qur'an as a manifest light (4:174) and asked us to follow it (7:157). The clear signs are there to bring us out of darkness into light (5:16; 14:1).

O mankind! Verily there hath come to you a convincing proof from your Lord; for We have sent unto you a light (that is) manifest. (4:174)

يَاأَيُّهَا النَّاسُ قَدْ جَاءَكُمْ بُرْهَانٌ مِنْ رَبِّكُمْ وَأَنْزَلْنَا إِلَيْكُمْ نُورًا مُبِينًا (١٧٤)

Surat al-Nisā'

"Those who follow the Messenger, the unlettered prophet, whom they find mentioned in their own (scriptures) — in the Law and the Gospel — for he commands them what is just and forbids them from what is bad (and impure); He releases them from their heavy burdens and from the yokes that are upon them. So it is those who believe in him, honor him, help him, and follow the light which is sent down with him. (7:157)

الَّذِينَ يَتَّبِعُونَ الرَّسُولَ النَّبِيَّ الْأُمِّيَّ الَّذِي يَجِدُونَهُ مَكْتُوبًا عِنْدَهُمْ فِي التَّوْرَاةِ وَالْإِنْجِيلِ يَأْمُرُهُمْ بِالْمَعْرُوفِ وَيَنْهَاهُمْ عَنِ الْمُنْكَرِ وَيُحِلُّ لَهُمُ الطَّيِّبَاتِ وَيُحَرِّمُ عَلَيْهِمُ الْخَبَائِثَ وَيَضَعُ عَنْهُمْ إِصْرَهُمْ وَالْأَغْلَالَ الَّتِي كَانَتْ عَلَيْهِمْ فَالَّذِينَ آمَنُوا بِهِ وَعَزَّرُوهُ وَنَصَرُوهُ وَاتَّبَعُوا النُّورَ الَّذِي أُنْزِلَ مَعَهُ أُولَئِكَ هُمُ الْمُفْلِحُونَ (١٥٧)

Surat al-A'rāf

There hath come to you from God a (new) light and a perspicuous book. (5:15)

قَدْ جَاءَكُمْ مِنَ اللَّهِ نُورٌ وَكِتَابٌ مُبِينٌ (١٥)

Surat al-Mā'idah

Alif Lām Rā. A book which We have revealed unto you in order that you might lead mankind out of the depths of darkness into light — by the leave of the Lord to the way of (Him) the Exalted in Power, Worthy of all Praise! (14:1)

الر كِتَابٌ أَنْزَلْنَاهُ إِلَيْكَ لِتُخْرِجَ النَّاسَ مِنَ الظُّلُمَاتِ إِلَى النُّورِ بِإِذْنِ رَبِّهِمْ إِلَى صِرَاطِ الْعَزِيزِ الْحَمِيدِ (١)

Surah Ibrāhim

Building Up of the Light Circuit in the Qur'an

God has created man in the best possible image (40:64) and made him His successor on Earth (2:30). As He is the Merciful, He did not leave him

Science Miracles: No Stick or Snakes

to live like an animal. He gave him a unique brain. After his creation, He asked the angels to prostrate to man in recognition of the struggle he would have to go through on earth. He promised to send man guidance, which he should expect and look for (20:123). For the nation of Muhammad, His light is contained in the Qur'an.

He chose Muhammad, who could neither read nor write (7:157), as a prophet. He was an orphan and helpless (93:6). He was among people who lived in a desert and who, in their ignorance, used to bury their daughters alive in the sand (81:8); however, they were very fond of their language and competed regularly in poetry. The Qur'an, being in their language, represented a great challenge to them. They could not produce ten *surahs* or even one like any part of the Qur'an. Those who attempted, failed. The Qur'an came in parts so it can be read slowly, carefully, and suiting the occasion (17:106). He made each *surah* to stand alone as guidance. The miracles in the Qur'an are quite different from those given to other prophets: No stick that transforms into a snake and no radiant hand. But the people of the Prophet Muhammad's time wondered why God did not send a miracle like one of those mentioned (6:37). God answered them on the spot in the next verse: All creatures that make a sound as they move on earth and birds that fly with wings are all nations like man (6:38). They were told that if they cannot appreciate the scientific miracles, they will be deaf, mute, and living in darkness (6:39). The Qur'an is full of miracles in the form of scientific allusions and may be classified under three main headings.

1. A language miracle to introduce the Qur'an to very keen competitive Arabs. It is also appreciated by the Arabic speaking peoples. There is no necessity to know any Arabic to appreciate its inimitability. Eighty percent of the Qur'an verses rhyme with three sounds made by four letters.
2. Guidance to the social sciences covering all aspects of life including economics, systems of taxation, law, and marriage.
3. Scientific allusions as signs or miracles. Some of these were mentioned in the previous chapters. They can be classified as follows:
   - Scientific allusions to historical events. They show the effect of natural phenomena. Some originated on Earth. Others may have been extraterrestial in nature. The signs are there not just to frighten Muhammad's nation but also to warn them that similar episodes may happen to us now.
   - Science related to the universe, man and his environment.

The light system in the Qur'an appears to be in the form of a circuit. It starts from God to the prophets and their holy books to man and then back to God carried by man. One can accept the light in the Qur'an metaphorically as guidance. On the other hand, it is possible to understand the light reaching man in a scientific manner. God's light is real and man is a complex electrical system.

From the very beginning, God, The Merciful (*Al-Rahman* الرحمن), established His light by writing the initial of His name, The Light (*Al-Nur* النور), at the beginning of the second *surah* to be revealed (68:1). At the same time, the letter *nūn* is the last letter of His name, *Al-Rahman*. He spread the *nūn* widely in the Qur'an to rhyme 3,123 verses (50.08%) with it. He placed the rest of His name *Al-Rahman*, الر (*alif lām rā*) then حم (*ḥā mīm*) in the middle of the Qur'an at the beginning of *surahs* 55–66 (chronologically). The Qur'an is described as a book of guidance and mercy (16:64). Both characters go hand in hand, He guides us as a gesture of His mercy.

One can imagine that His name, *Al-Rahman*, is the beginning of the light circuit for guidance. The letters of the name will seek each other for completion. *Alif lām rā* (الر) will seek the *ḥā mīm* (حم) and the five letters will connect with 3,123 *nūn*s, placed at the end of the verses. This is not unlike the flow of an electric light current. The name will make a network connecting with the other beautiful names of God to create the systems of guidance which have been described in the previous chapters. The appearance of God's name, *Al-Rahman*, in the manner described above, appears as if He signed His name on the whole of the Qur'an using the symbolic letters. If one imagines how the Qur'an could be stored in *umm al-kitāb* one would appreciate that it cannot be stored in the way we have the Qur'an in our hand. It is probably stored by a method far more sophisticated than how we store our information on a computer. God created most things in a circular or ball shaped fashion like the Earth, the Sun, the Moon, or the planets. One can see and imagine how the light path was made according to revelation. The *nūn* was put first to mark the light terminals, the *alif lām rā*, some time later, then the *ḥā mīm* later still, thus completing the light path in the Qur'an and the first part of the circuit. Figure 18 is a model to illustrate how the light moves through God's name, *alif lām rā*, to *ḥā mīm* then to the *nūn*s. Figures 19 and 20 show the proportionate distribution of verses rhyming with *nūn* and their area. The radar charts, Figures 21 and 22, are designed to show changes in, or frequencies of, data relative to a center point and each other. Each has its own value axis radiating from the center point. Lines connect all the data markers in the same series. Figure 21 shows the proportionate distribution of verses rhyming with *nūn*, as shown

Science Miracles: No Stick or Snakes 115

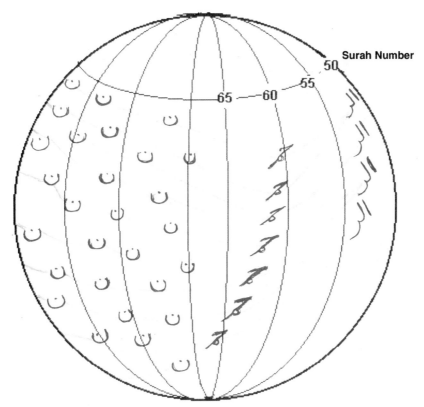

Figure 18: Model illustrating the flow of God's name, The Merciful, Al-Rahman (الرحمن) through the Qur'an.

in Figure 19, but now they are plotted on a radar graph together with the *alif lām rā* and *ḥā mīm* in cronological order of the *surah*s as they were revealed in the Qur'an. Please note the *alif lām rā* are present at the base of two feeder systems in the shape of two triangles. Each is a scalene triangle but they have equal bases and coincide completely. The two feeder triangles of *alif lām rā* are directed to the central point where all the *Ha Mim*s accumulate. Then it spreads1 out to the *nūn*s. If we repeat Figure 21, however, replacing the proportional distribution of the verses rhyming with *nūn* by their order of revelation, Figure 22 shows a great difference. The *alif lām rā ḥā mīm* are now coinciding at the central point and also coinciding with almost all the *nūn*s at the center except for a few scattered points. Figure 22 shows scientifically, by radar chart, that God's name, *Al-Rahman*, flows with the letters that make the name through and across a unique path in the Qur'an.

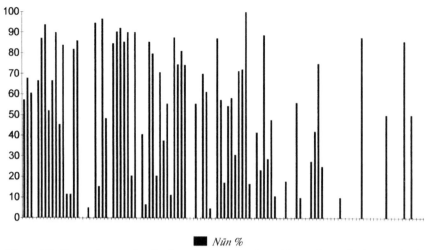

Figure 19: *Proportionate distribution of verses rhyming with* nūn *in* surahs *in cronological order in the the Qur'an.*

No light system or circuit is complete unless it returns to the originator. The system of light returning to God in the Qur'an is complex. One can accept it metaphorically, however, it may also be explained in scientific terms. God is The Imager (*Al-Muṣawwir* المصور). Primary images are always reversed. This may explain why *ṭa sīn mīm* (طسم) looks reversed. It may also indicate the direction of the light path as it returns to God.

The light is provided in the Qur'an by The Guide (*Al-Hādī* الهادى), by sending prophets, where they are needed, in places of injustice on Earth. The prophets are suitably chosen, given the most suitable miracles and signs to guide people before any punishment befalls them. This is the prin-

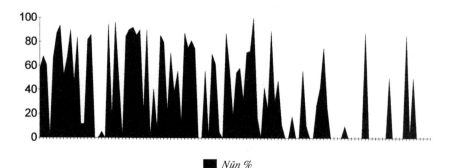

Figure 20: *Proportionate distribution by area of verses rhyming with* nūn *in* surahs *in cronological order in the Qur'an.*

Science Miracles: No Stick or Snakes

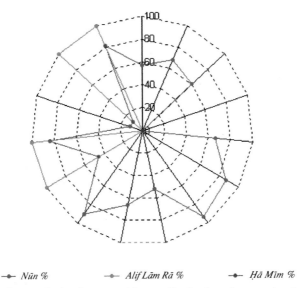

— Nūn %   — Alif Lām Rā %   — Ḥā Mīm %

Figure 21: *Radar graph showing proportionate distribution of verses rhyming with* nūn, *also* alif lām rā *and* ḥā mīm *in* surahs *ordered cronologically. Please observe that* alif lām rā *is delivered from two triangular feeder systems coming to the central point, where the* Ḥa Mims *accummulate, then spreading to the* nūns.

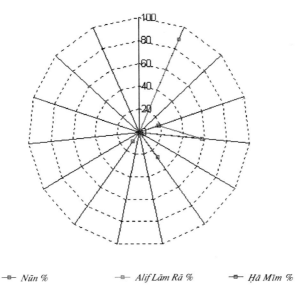

— Nūn %   — Alif Lām Rā %   — Ḥā Mīm %

Figure 22: *A repeat of Figure 21 but with the proportionate distribution of the* surahs *rhyming with* nūn *in cronological order, also* alif lām rā *and* ḥā mīm *point indicating a common and straight path.*

ciple of The Equitable (*Al-Muqsit* المقسط). The guiding system uses the adjectives and characters of the beautiful names of God. They work their way through all the Qur'an to educate those who read it — man, jinn, or otherwise. Those who read the Qur'an will receive such light, it will be recorded in their brain, and it will affect them in this life and in the life to come.

The Spread of Light into Man

As mentioned earlier, the human body is a complex electrical machine. The primary focus of electricity starts in the heart and spreads to the rest of the body through another complex system of ion exchange controlled by sodium and potassium ions, creating a continuous electrical current. All of the body functions in this way. As explained earlier, brain acivities (thoughts?) are measured by the EEG (electroencephologram). We have seen the recording of excessive stimulation in the brain in epilepsy. Thus no one should be surprised at thoughts being recorded. We are required to read the Qur'an with great thought. The process of thinking will produce an impulse in the human brain. The thought is the outcome of guidance given by the Qur'anic light circuit. The impulses will be stored to make a frame of reference of background knowledge.

As a result of the effect of guidance man will begin to act. If he thinks adversely and does not act, there will be no score. If he acts, it will only be scored as minus one point (6:160). On the other hand if his electric impulses were based on sound guidelines from the Qur'an, his actions will be constructive. Every good deed will be scored for him more than double (57:11). Some of these actions may be of such great value that they may score ten or seven hundred times their worth (6:160; 2:261).

Surah Qāf (ق), which starts with this symbolic letter, makes many scientific allusions. One of them is that God has created two receivers to receive what we say or do, stationed on our right and left (50:17–18). He does not say precisely where these two receivers exist. As has been explained elsewhere (Abbas 1997), they may actually be present in the brain in one of the numerous centers still unknown to man and programmed by God. God says that each of us has something that flies attached to the neck, and a book will be presented to him on the Day of Judgment (17:13). This book may be interpreted as a computer printout; it is not difficult for us to imagine that the scoring system in man is built from light signals scored as a result of light impulses of our thoughts and deeds.

Science Miracles: No Stick or Snakes

The previous detailed explanation shows that thoughts built on education and guidance from the Qur'an followed by action is not unlike the previous example of a person who runs from a danger which he saw or heard. The difference is that the thoughts and deeds based on Qur'anic education are part of a circuit where they are recorded in the brain and will be presented to man's Creator.

Behold, two (guardian angels) appointed to learn (his doings) learn (and note them), one sitting on the right and one on the left. Not a word does he utter but there is a sentinel by him, ready (to note it). (50:17–18)	إِذْ يَتَلَقَّى الْمُتَلَقِّيَانِ عَنِ الْيَمِينِ وَعَنِ الشِّمَالِ قَعِيدٌ (١٧) مَـــا يَلْفِظُ مِنْ قَوْلٍ إِلاَّ لَدَيْهِ رَقِيبٌ عَتِيدٌ (١٨) Surah Qāf
Every man's fate We have fastened on his own neck: On the Day of Judgment We shall bring out for him a scroll, which he will see spread open. (17:13)	وَكُلَّ إِنْسَانٍ أَلْزَمْنَاهُ طَائِرَهُ فِي عُنُقِهِ وَنُخْرِجُ لَهُ يَوْمَ الْقِيَامَـــةِ كِتَابًا يَلْقَاهُ مَنْشُورًا (١٣) Surat al-Isrā'

This scientific analysis may explain what God says in the Qur'an: If He does not assign light for us we will live without any light (24:40). When He opens our heart to the guidelines of Islam, we will be on the light line or circuit from God (39:22). This can only be affected by acting on the stored information we gain from our education from The Guide, *Al-Hadi,* in the Qur'an (57:28).

Or (the unbelievers' state) is like the depths of darkness in a vast deep ocean, overwhelmed with billow topped by billow, topped by (dark) clouds: Depths of darkenss, one above another: if a man stretches out his hand, He can hardly see it! For any to whom God gives not light, there is no light! (24:40)	أَوْ كَظُلُمَاتٍ فِي بَحْرٍ لُجِّيٍّ يَغْشَاهُ مَوْجٌ مِنْ فَوْقِهِ مَوْجٌ مِـــنْ فَوْقِهِ سَحَابٌ ظُلُمَاتٌ بَعْضُهَا فَوْقَ بَعْضٍ إِذَا أَخْرَجَ يَدَهُ لَـــمْ يَكَدْ يَرَاهَا وَمَنْ لَمْ يَجْعَلِ اللَّهُ لَهُ نُورًا فَمَا لَهُ مِنْ نُورٍ (٤٠) Surat al-Nūr
Is one whose heart God has opened to Islam, so that he has received enlightenment from God, (no better than one hard-	أَفَمَنْ شَرَحَ اللَّهُ صَدْرَهُ لِلإِسْلاَمِ فَهُوَ عَلَى نُورٍ مِنْ رَبِّهِ فَوَيْـــلٌ لِلْقَاسِيَةِ قُلُوبُهُمْ مِنْ ذِكْرِ اللَّهِ أُولَئِكَ فِي ضَلاَلٍ مُبِينٍ (٢٢) Surat al-Zumar

hearted)? Woe to those whose hearts are hardened against celbrating the praises of God! They are manifestly wandering (in error)! (39:22)

O you that believe! Fear God, and believe in His Messenger, and He will bestow on you a double portion of His mercy: He will provide for you a light by which you shall walk (straight in your path), and He will forgive you (your past): For God is Oft-Forgiving, Most-Merciful. (57:28)

يَاأَيُّهَا الَّذِينَ آمَنُوا اتَّقُوا اللَّهَ وَآمِنُوا بِرَسُولِهِ يُؤْتِكُمْ كِفْلَيْنِ مِنْ رَحْمَتِهِ وَيَجْعَلْ لَكُمْ نُورًا تَمْشُونَ بِهِ وَيَغْفِرْ لَكُمْ وَاللَّـهُ غَفُورٌ رَحِيمٌ (٢٨)

Surat al-Ḥadîd

Completion of the Light Circuit

There are many people who serve unjust causes and try to interrupt God's circuit of light. Their attempts end in failure. God refuses to allow His light circuit to break. He ensures its continuity (9:32; 61:8). All His effort to send messengers (33:43) with guidance and signs, aims at getting us out of darkness into His Light (2:257; 57:9; 65:11). The light circuit is not truly complete unless those who carry it while it is shining around them take it back to God. They will have a high score of light points which will permit them to enter Paradise (6:122; 57:12).

Fain would they extinguish God's light with their mouths, but God will not allow but that His light should be perfected, even though the unbelievers may detest (it). (9:32)

يُرِيدُونَ أَنْ يُطْفِئُوا نُورَ اللَّهِ بِأَفْوَاهِهِمْ وَيَأْبَى اللَّهُ إِلاَّ أَنْ يُتِمَّ نُورَهُ وَلَوْ كَرِهَ الْكَافِرُونَ (٣٢)

Surat al-Tawbah

Their intention is to extinguish God's light (by blowing) with their mouths: But God will complete (the revelation of) His light, even though the unbelievers may detest (it). (61:8)

يُرِيدُونَ لِيُطْفِئُوا نُورَ اللَّهِ بِأَفْوَاهِهِمْ وَاللَّهُ مُتِمُّ نُورِهِ وَلَوْ كَرِهَ الْكَافِرُونَ (٨)

Surat al-Ṣaf

He it is who sends blessings on you, as do His angels, that He may bring you out from the depths of darkness into light: And He is full of mercy to the believers. (33:43)

هُوَ الَّذِي يُصَلِّي عَلَيْكُمْ وَمَلاَئِكَتُهُ لِيُخْرِجَكُمْ مِنَ الظُّلُمَاتِ إِلَى النُّورِ وَكَانَ بِالْمُؤْمِنِينَ رَحِيمًا (٤٣)

Surat al-Aḥzāf

Science Miracles: No Stick or Snakes

God is the protector of those who have faith: From the depths of darkness He will lead them forth into light. Of those who reject faith the patrons are the evil ones: From light they will lead them forth into the depths of darkness. They will be companions of the fire, to dwell therein (forever). (2:257)

اللّهُ وَلِيُّ الَّذِينَ آمَنُوا يُخْرِجُهُم مِّنَ الظُّلُمَاتِ إِلَى النُّورِ وَالَّذِينَ كَفَرُوا أَوْلِيَاؤُهُمُ الطَّاغُوتُ يُخْرِجُونَهُم مِّنَ النُّورِ إِلَى الظُّلُمَاتِ أُوْلَئِكَ أَصْحَابُ النَّارِ هُمْ فِيهَا خَالِدُونَ (٢٥٧)

Surat al-Baqarah

He is the one who sends to His servant manifest signs, that He may lead you from the depths of darkness into the light. And verily, God is to you Most-Kind and Merciful. (57:9)

هُوَ الَّذِي يُنَزِّلُ عَلَى عَبْدِهِ آيَاتٍ بَيِّنَاتٍ لِيُخْرِجَكُم مِّنَ الظُّلُمَاتِ إِلَى النُّورِ وَإِنَّ اللَّهَ بِكُمْ لَرَؤُوفٌ رَحِيمٌ (٩)

Surat al-Ḥadîd

A messenger, who rehearses to you the signs of God containing clear explanations, that he may lead forth those who believe and do righteous deeds from the depths of darkness into light. And those who believe in God and work righteousness, He will admit to Gardens beneath which rivers flow, to dwell therein forever: God has indeed granted for them a most excellent provision. (65:11)

رَسُولاً يَتْلُو عَلَيْكُمْ آيَاتِ اللَّهِ مُبَيِّنَاتٍ لِيُخْرِجَ الَّذِينَ آمَنُوا وَعَمِلُوا الصَّالِحَاتِ مِنَ الظُّلُمَاتِ إِلَى النُّورِ وَمَن يُؤْمِن بِاللَّهِ وَيَعْمَلْ صَالِحاً يُدْخِلْهُ جَنَّاتٍ تَجْرِي مِن تَحْتِهَا الْأَنْهَارُ خَالِدِينَ فِيهَا أَبَداً قَدْ أَحْسَنَ اللَّهُ لَهُ رِزْقاً (١١)

Surat al-Ṭalāq

Can he who was dead, to whom We gave life, and a light whereby he can walk among men, be like him who is in the depths of darkness, from which he can never come out? Thus to those without faith their own deeds seem pleasing. (6:122)

أَوَ مَن كَانَ مَيْتاً فَأَحْيَيْنَاهُ وَجَعَلْنَا لَهُ نُوراً يَمْشِي بِهِ فِي النَّاسِ كَمَن مَّثَلُهُ فِي الظُّلُمَاتِ لَيْسَ بِخَارِجٍ مِّنْهَا كَذَلِكَ زُيِّنَ لِلْكَافِرِينَ مَا كَانُوا يَعْمَلُونَ (١٢٢)

Surat al-An'ām

One day shall you see the believing men and the believing women — how their light runs forward before them and by their right hands: (Their greeting will be): "Good news for

يَوْمَ تَرَى الْمُؤْمِنِينَ وَالْمُؤْمِنَاتِ يَسْعَى نُورُهُم بَيْنَ أَيْدِيهِمْ وَبِأَيْمَانِهِم بُشْرَاكُمُ الْيَوْمَ جَنَّاتٌ تَجْرِي مِن تَحْتِهَا الْأَنْهَارُ خَالِدِينَ فِيهَا ذَلِكَ هُوَ الْفَوْزُ الْعَظِيمُ (١٢)

Surat al-Ḥadîd

you this day! Gardens beneath which flow rivers to dwell therein for aye! This is indeed the highest achievement! (57:12)

Light Points Can Only Be Gained on Earth

The hypocrites, those who mislead others claiming that they are virtuous, certainly cannot deceive God. Their light scores will be so low that it will lead them to Hell. In the next life they will try to borrow, in vain, extra light points from the believers on the pretence that they were like them. Light points can only be obtained on Earth (57:13). God confirms that He will complete His light (9:32; 61:8).

One day will the hypocrites — men and women — say to the believers: "Wait for us! Let us borrow (a light) from your light!" It will be said: "Turn you back to your rear! Then seek a light (where you can)!" So a wall will be put up between them, with a gate therein. Within it will be mercy throughout, and without it, all alongside will be (wrath and) punishment! (57:13)

يَوْمَ يَقُولُ الْمُنَافِقُونَ وَالْمُنَافِقَاتُ لِلَّذِينَ آمَنُوا انظُرُونَا نَقْتَبِسْ مِنْ نُورِكُمْ قِيلَ ارْجِعُوا وَرَاءَكُمْ فَالْتَمِسُوا نُورًا فَضُرِبَ بَيْنَهُمْ بِسُورٍ لَهُ بَابٌ بَاطِنُهُ فِيهِ الرَّحْمَةُ وَظَاهِرُهُ مِنْ قِبَلِهِ الْعَذَابُ (١٣)

Surat al-Ḥadîd

The Symbolic Letters Are Initiators and Seal of the Light Circuit

From what was mentioned earlier, the symbolic letters seem to initiate, propagate, confirm, and seal the path of the light circuit in the Qur'an. The *nūn* initiates the symbolic language and confirmed the light of the Qur'an. It forms the base and foundation of God's name, *Al-Rahman*, the pillar on which guidance relies. The Qur'an's objective is to guide and be a mercy to mankind. The symbolic letters mark the introduction of guidance by prophets who were asked to be patient in delivering their message. The letters forewarn of the universe's natural phenomena, make allusion to science, indicate the reception and recording of man's actions, and ultimately the return of light to God with those who are worthy of Paradise.

Chapter Fifteen

Heaven's Library

Whoever recites the Qur'an is struck by the endless number of books that are mentioned in connection with the creation, the universe, the heavens, the earth and what is in between, and the creatures, including man. These books can be summarized as follows:

Umm al-Kitāb

God states that the Qur'an contains confirmed verses, which are *umm al-kitāb* and other verses, which are allegorical (6:7). *Umm al-kitāb* is a high and wise book possessed by God (43:4). He blots out or confirms what He wills (13:39). He gave several prophets holy books. The Torah is the book given to the Prophet Moses (6:91; 14:5) and the *Injil* was given to the Prophet Jesus (7:157). God always refers to these books as "The Book." He named the book that He gave to the Prophet David the *Zabur* (4:163; 17:55). God told the Prophet Yahya (John) to hold the book firmly (19:12). He refers to the books of earlier prophets as *suhuf al-ula* (87:18).

Al-Lawḥ Al-Maḥfūẓ (اللوح المحفوظ) or the Preserved Tablet

In this book, God records those and their blood relations who emigrated and fought with the Prophet Muhammad (8:75). He also records in it that the year is twelve months when He created the heavens and earth (9:36). This book also records everybody who lives and dies until the Day of Doom (30:56).

Books Related to the Universe and Earth

1. God does not miss any quantity, even if by only an atom's weight, in the heavens or in earth, or anything smaller or bigger than that; and He records it all in a manifest book (34:3; 10:66).
2. He has the keys to the unknown and He knows what is in the earth and what is in the sea. There is no leaf that drops except that He is aware of; there is no seed, in the darkness of earth, in humid or dry soil, except it is in a manifest book (6:59).
3. Every creature that creates sound on Earth as it moves is provided for by God. He knows where it settles and where its store is and each is in a manifest book (11:6).
4. There is nothing in the heavens or earth, missed or unseen, but it is in a manifest book (22:70; 27:75).
5. God knows all catastrophes that happen in heaven or earth beforehand and it is recorded in a book (57:22).
6. Every village will be destroyed or punished before the Day of Judgment and this is recorded in a book (17:58).

Books for Statistics

There will be a book where everything big or small is counted (18:49). Another book of statistics keeps a record of the people whom God resurrects and what they have done and what they have left behind (36:12). God has kept everything statistically in a book (78:29).

Books for Nations and Books for Individuals

There is a book for every nation (45:28).

It seems that God has more than one book for each individual.
1. He has a book that talks in truth about each of us (45:29).
2. Everybody will have a kind of flying object fastened to his neck (fate or destiny according to some translators) and he will be presented with a book on the Day of Judgment (17:13).
3. Prayers are in a book for the believers (8:75; 10:61; 83:18,20).
4. The death of every soul is in a book (3:145).
5. Everybody will be presented with his own book (17:71; 69:19; 69:20; 69:25; 84:7; 84:10).

6. The books of the wicked are placed in *Sijjīn* and they are numbered (83:7; 83:9). Books for the good are in *'Illiyīn* and they are numbered (83:18,20).
7. Catastrophes that befall man are known to God in advance and are in a book (57:22).
8. God created us from dust and then from sperm and made us in pairs. He knows of every pregnancy or evacuation of the uterus, and He knows everybody who stays alive for a long time or whose life is curtailed, all this is in a book (35:11).
9. Last but not least, God is aware how much the Earth's weight will be reduced when one of us dies. This indicates that He can measure our spirit (*rūḥ*) and He records each in a book well-preserved (50:4).

The Size of These Books

The immensity of the information stored in these books surpasses everybody's imagination. How can God keep all this — collate, classify, file, and store all these books and all the information they contain? How can He then access this great volume of information stored? How can He identify the information of the items mentioned in each of the books and how can He find who said what or, who thought what on a certain day of a certain year at the precise second. I recall, when I was a child, asking my father, "How did God keep all these books and who looks after them?" His simple answer was, "God has a great big library and many angels to look after it." My questions were in the 1930s and 1940s long before computers. I have always imagined that all this information must be in billions upon billions upon billions of books written on an endless number of sheets of paper. I thought it would take years to find any piece of information and I thought that the Day of Judgment would be endless. Yet I knew all the time that God, if He wanted anything, only has to say, "Be," and it is. I have always wondered how God is going to do that. It is only recently through the discovery of computers, hard disks, floppy disks, and compact disks that we have developed the power to store great amounts of information. It should not, therefore, be difficult for us to appreciate that God has a system faster and more efficient than any known to mankind. It must be billions of times more powerful than any human computer. The heavenly library must have an unbelievable power for collecting, collating, filing, storing, and retrieving every minute detail of information present in each of these books.

The Symbolic Letters and Their Possible Role as Computer Markers

We suggest that the Glorious Qur'an is provided with a light system initiated, propagated, confirmed and sealed by the symbolic letters. They sign the name of God, probably in light, across and through the Qur'an, going round in a circular motion to connect with the letter *nūn* at the end of 50.08 percent of the verses of the Qur'an. We suggest that the symbolic letters are possibly there to act like computer markers, distinguishing the Qur'an in *umm al-kitāb* with God's name signed, across and through it, in glorious light. This would make the Qur'an stand out in *umm al-kitāb* among all other books.

CHAPTER SIXTEEN

CONCLUSION

It is not necessary to know Arabic to recognize that the Qur'an is a linguistic miracle. It is a large text composed of 330,733 letters, making 77,845 words within into 6,236 *ayāt* (verses), divided into 114 *surah*s (chapters). More than half of the verses (50.08 percent) rhyme with the letter *nūn* (ن). Another 30 percent rhyme with either *mīm* (م), *alif* (ا), or *yā* (ى). The *alif* and *yā* sound the same and are used interchangeably according to orthographical rules. The Qur'an also contains, besides guidance and faith, social sciences like observations on marriage, divorce, law (both civil and criminal), contracts, and testimonials. It also contains physical sciences like the creation of the universe, plate tectonics, water and estuarine cycles, the fourth dimension, human embryology, and psychology. No book written by a human being with such a wide range of subjects can follow such a rigid rhyming scheme.

The 6,236 verses of the Qur'an are threaded together by loose rhymes into short or long sequences. The rhythms of these sequences vary sensibly according to the subject matter. It swings from the steady march of the straight forward social and physical sciences to the majesty of God — the eminence of the Day of Judgment, the torments and terror of Hell and the joys and delights of Paradise. This is all done in great splendor, drama, and beauty.

Twenty-nine *surah*s begin with one to five individual letters, which are referred to as *al-Muqatta'āt* or parts of words. They look simple, but they are the building blocks of the Glorious Qur'an. God challenges the Arabs, who are keen about their language, to imitate any part of it. This book attempts to explain these letters. They are termed the "symbolic letters" because nobody seems to know what they mean. Our explanation of them is based on their relationship to the beautiful names of God, thus ensuring a safe and firm ground. To facilitate an analysis of the Qur'an, the *surah*s are considered in cronological order of revelation.

As shown earlier, the letter *nūn* (ن) is the beginning of two of God's names: The Light (*Al-Nur* النور) and The Beneficent. If the *nūn* at the beginning of verse 68:1 (second *surah* by cronology) is replaced by His two names, the verse reads as follows: "Beneficial light is what the pen can produce in continuous lines." It has been shown that the *nūn* also forms a wide base and foundation as the end letter of God's name *Al-Rahman*. The remaining five letters are given at the beginning of *surahs* 50–66 by revelation, where the *alif lām rā ḥā mīm* (الرحم) seem to be there to connect with the *nūn*s through and across the whole of the Qur'an, supporting it with light and guidance.

The letter *qāf* (ق) seems to mark the scientific verses present in the *surah* of the same name. In this *surah*, God describes part of the creation of the universe, the fourth dimension, extension of the *rawasi* or lithosphere, creation of man from a selected strain of mud and a possible weight for our spirit or *rūḥ*. He is nearer to us than the "venus return" to our heart. He has two receivers set on the right and the left, probably in our brain, programmed to record our thoughts and deeds. It describes the food cycle and how He rehydrates land to create life and possibly man to resurrect him. It refers to the creation of life and death by supersonic waves and it also refers to the use of the universe's natural phenomena as a method for selected mass destruction. It refers to science beyond time that may throw some light on our psychological behavior and also on our nature.

The Patient is one of God's names. Though not mentioned as such in the Qur'an, it is understood by inference. This is perhaps why God put the letter *ṣād*, the initial of this name, at the beginning of a *surah* that infers God as the symbol of patience. He was patient with all the people who went out of their way to destroy the path to Him. He showed them signs, sent them messengers to teach, preach and warn before He finally punished them. He was also patient with the Prophet David and the Prophet Solomon. Both forgot their duties because of a woman. He still forgave them and was patient with them. His third example was the Devil, He could have destroyed him for disobeying Him, instead he gave him respite until the Day of Resurrection. He gave the Prophet Muhammad a human symbol of patience, the Prophet Job. The Prophet Muhammad had a very difficult job, he was given a message full of science and knowledge while he could not read or write himself. He needed to be very patient.

The letters *ṭā hā* are traditionally understood as one of the names of the Prophet Muhammad. They begin *surah* 20 — most of which is concerned with the Prophets Moses and Adam. The letter *ṭā* appears at the beginning of another three *surah*s that are also concerned with the Prophet Moses and one of them, *surah* 28, seems to complete the narration of the Prophet Moses that began in *surah* 20. The letter *hā* is the beginning of God's name, *Al-Hādī*, The Guide (الهادى). God's system of guidance has relied on sending prophets to investigate and offer solutions for a epidemiological problems. *Surah*s 20 and 28 answer the five questions beginning with "W" concerning a prophet. The Prophet Moses is used as an example. *Where* and *why* was a prophet needed? Egypt was the place that needed a prophet as pharaoh claimed himself a god, allowed the killing of the male children of the Children of Israel. *Who* can be a prophet? The Prophet Moses was chosen. He was given signs in the form of miracles; God supported him by removing his stutter and making his brother a prophet to help him. When God thought it opportune, Moses was given the task of spreading God's guidance to Pharaoh, his soldiers, his administration, and the people of Israel. By *what* technique did he do this? He did it by preaching, guiding, performing miracles, the exodus, more miracles and finally the holy book, the Torah. God promised to send His guidance through prophets ever since Adam inhabited the Earth.

Surah Yā Sīn (يس), which has a high mystical value among Muslims, is recognized as another name for the Prophet Muhammad. The *surah* that has the same name describes God's power to give life and to cause death. It also states that He is the peace that everyone seeks.

A computer search for *alif lām mīm ṣad* (المص) shows only nine words in the Qur'an which begin with these letters. The only one of significance is God's name, The Imager (*Al-Muṣawwir* المصور). The *surah* beginning with these letters contains three verses that show inimitable scientific foresight. He "images" man in the womb. Previously, we gave Magnetic Resonance Imaging, the fact that one-half of the body is a mirror image of the other, the genetic map, and the Human Genome as evidence. An interesting metaphor is human time travel transfer, which has been used in science fiction, where people are "beamed" from the present to the future and from Earth to a spacecraft. God "beamed" us all up to testify that He is our Creator. How we appeared before we existed is a question that can only be answered by The Imager. The wife of Adam was created from him. The process of budding is known in yeasts and plants. It could only have been done in humans by the master of human genetics — God.

The five letters at the beginning of *surah* 19 represent fourteen names of God that the Prophet Zachariah used to request God to give him a child to carry on his message. Examples of such prayer are given in the text. The *surah* shows God's power of creation by the regeneration of senile organs, as in the case of the Prophet Zachariah's wife, who was old and barren. God had already given the Prophet Abraham, who was in a similar situation, two children who became prophets. God gave the unique name *Yahya* يحيى to the Prophet Zachariah's son. The name comes from God's name, The Giver of Life (*al-Muhyi* المحي). The spelling of the verb *yuhiya* to the name of the Prophet they are pronounced differently. Having children after menopause is now possible, so it should not be difficult for God. Parthenogenesis, reproduction without sexual union, is happening all the time among some birds, reptiles, insects and plants. The Prophet Jesus was created by a process reminiscent of parthenogenesis. Cysts on the ovaries, by the name of teratomas or dermoid, are occurring in women all the time. They contain teeth, hair, or other structures like the lung or intestine. These cysts do not migrate to the uterus, and if they do, they do not continue to develop into a child. A report has recently appeared in which the white blood cells of a child possessed chromosomes from his mother only. The Prophet Jesus and the Prophet Yahya were created by a similar process through an act of God, so why should one be called the "Son of God" and not the other. Both were born, both will die, and both will be resurrected. God does not need a son to look after His interests, unlike the prophets Zachariah or Abraham. God is Eternal.

The letters *ṭā sīn mīm* (طسم) are present at the beginning of the *surah*s 26, 27 and 28. The *mīm* is missing in *surah* 27. Verse 2 of the three *surah*s refers to the signs of the Qur'an. Careful study shows that the *ṭā* is probably there to show the relation among *surah*s 20, 26, 27, and 28. They share the story of the Prophet Moses. The *ṭa* refers to the sacred mountain *ṭūr*. *Ṭūr* connects *surah*s 20 and 28 to illustrate the system of guidance through the prophets, using the Prophet Moses as an example. If one reverses the three letters *ṭā sīn mīm* to read *mīm sīn ṭā* (مسط), it will be the correct beginning of the word *masṭūr*, which is mentioned in *ayat*s 52:2 and 68:1 (in its verb form). Adding the letter *qāf* after the *mīm* yields God's name Al-Muqsiṭ (مقسط, The Equitable), which is the theme of the three *surah*s, 26, 27, and 28. The word *mastūr* in *ayah* 68:1 marks the inimitability of the Qur'an by rhyming more than half of its verses with the letter *nūn* (ن). These dynamics are controlled by the symbolic letters.

God's name, The Equitable, prevails on the three *surah*s. He does not punish a people who have committed definite crimes against the Earth and those on it without first sending them a prophet with signs and a system of guidance. Eventually, he punishes them by a selective method of mass destruction in which He saves the worthy. This happened with almost all His prophets. The *surah*s also refer to the signs of the Qur'an which are no longer like those given to the prophets Moses or Jesus. The signs or miracles in the Qur'an are in the form of numerous scientific allusions. The issues of the social sciences are mentioned in the Qur'an, as are the subjects of scientific inquiry, such as the creation of the lithosphere, the presence of the estuarine cycle, the stages of the Moon for celestial navigation, and he reserves the power of wind to himself. People have been wondering when their punishment will come: God hinted that such punishment may have

Science Miracles: No Stick or Snakes 131

already begun, perhaps by means of one of the universe's natural phenomena.

God used several of the universe's natural phenomena for selective mass destruction. Noah lived 950 years and he manufactured a boat with vision and direction of his thoughts from God. Perhaps the purpose of his long life was to allow the natural phenomena to take place and to give him time to build the ark. He described the flood, which He used as punishment, as *far al tanūr,* which means "overflow of the face of the earth." The sea level rose to an exceptionally high level and was covered with mountainous waves. This can be explained by the melting of the ice caps during Noah's lifetime. The astronomical theory of climate change shows by mathematics that during the Earth's orbit of the Sun it gets nearer to the Sun every 100,000 years by about 10 million km. It also comes nearer to the Sun every 40,000 years as it rotates on its axis from less than 22 degrees to more than 24 degrees. It also wobbles, as it spins, every 22,000 years. This affects the amount of radiation falling on the poles. Geological studies have proven this theory by studying the concentration of heavy and light oxygen molecules in the shells of the crustaceans foraminifera. The ocean water contains both oxygen molecules. When evaporated the light molecules will go into the clouds and during the ice age they will drop as snow on the ice caps. This means there will be a greater abundance of the heavy oxygen molecules in the ocean during the ice age. The shells of foraminifera will act as a tape recorder of what happened in the ocean. When the variation is plotted it corresponds to the mathematical theory. In 1976, three scientists plotted the variation created by the falling of sediments to the sea bed and how they align themselves to the Earth's magnetic field. The magnetic field "flips" into reverse direction at different intervals. Their results indicate that it "flipped" 700,000 years ago.

Concerning the melting of the ice caps, it could be due to a meteorite impact on the ice cap or in the ocean. Another possibility is the occurrence of tsunami or seismic sea waves. This is a gigantic sea wave about 50 meters high that suddenly appears at the coast. An earthquake at the bottom of the sea generates a very strong wave that travels along the bottom of the sea until it reaches the coast. Warning systems are now present in the Pacific to predict their occurrence. Alexandria, Egypt, was leveled to the ground by a tsunami in 365 AD. Such waves can travel 4.5 km inland.

The people of Lot woke up to a horrendous blast or cry. Their city was turned upside down. A rain of stones, baked in hell, fell on them layer upon layer. Each stone was labelled to hit a specific person. A similar ordeal overtook the Companions of the Elephant, who came to destroy the *Ka'ba* with their elephant. God sent flocks of fliers or birds that threw on them hellish stones. When the attack had finished, the place was empty and looked as if it was strewn with the dirty remains of stalks and straw eaten and stepped on by cattle.

The picture of the destruction of Lot's people is very similar to an impact on the Earth by a meteorite or asteroid during which earth evacuates leaving a crater. The shattered earth will come back in flames, smoke, and fire. An impact that took place 65 million years ago at Chicxulub, on the Yucatan peninsula off the Gulf of Mexico, was thought to have been the cause of the extinction of many animals, like the dinosaur, and many plants as well. Impact craters are present throughout the world. Nordlingen in southern Germany is built in an impact crater. Meteorites could have been the objects that fell on the Companions of the Elephant.

The people of 'Ād saw a cloud approaching their valley. Pleased because they thought it to be a rain cloud, they found instead a devastating wind, accompanied by cold, a tremendous screaming sound, excessive violence, and lightening bolts. It stayed with them for seven nights and eight days. The wind plucked out the people as though they were palm trees torn from their roots. They lay in its path prostrate; there were no survivors; everything was destroyed. Hurricanes and tornadoes are similar to the wind described above; however, they do not last as long and they are usually accompanied by rain.

In the early morning, the people of Thamūd experienced an earthquake accompanied by a blast and a lightening bolt. The morning found them prostrate in their homes like dry stubble used by herdsman to make pens for the enclosure of cattle. Today, earthquakes are known to occur in the area of northwest Arabia between Medina and Syria where the people of Thamūd lived.

Before morning, the Companions of the Wood felt an earthquake and a blast. Their city was overshadowed by a cloud, which ushered in a day of great torture. They were left prostrate in their homes. This event may have occurred as a result of seismic activity accompanied by a mushroom cloud from a volcanic eruption; however, this catastrophe could have been caused by a meteorite because fire and smoke have been reported as a result of meteorite impacts.

God asked the Prophet Moses to strike the sea with his rod. It divided the sea and each part became like a huge mountain of water. Pharaoh and his soldiers chased them in insolence and spite. When the water caught up with Pharaoh and he was about to drown, he declared that he recognized the God of Israel. The traditional explanation is that the sea was split into twelve passages one for each of the tribes of Israel. The sides of each passage were like a big mountain. None of those who passed safely through had a wet saddle or wet clothes.

Presumably, the Prophet Moses was heading toward Sinai and the Gulf of Suez. The Red Sea is a tidal sea, if a family of strong tornadoes, arising from cumulonimbus clouds right above them developed, moving along the north-easterly prevailing wind, they would probably be able to create the passages. The funnels of tornadoes have a suction power of up to 200 mph, which could easily make several clear passages in the water for the twelve tribes of Israel. Each tornado with its suction power would create directly under it a mountain of water as mentioned in the Qur'anic description. The continuous suction would create a passage on either side of the water mountain. Two additional tornadoes suitably distanced on either side of the first would make two absolutely clear passages of land for Moses' followers to use. The twelve passages would, theoretically, be made by a family of thirteen tornadoes.

As the tornadoes leave the water and travel overland, they would become slower, losing their power and dropping the mountains of water. These would come down on Pharaoh's soldiers who were chasing Moses's followers, who would have already crossed.

God has chosen man to be His successor on Earth. We should protect ourselves against high winds, earthquakes, bolts of lightening, floods, tsunamis, etc. Our biggest challenge is objects from outer space. Earth-based observatory stations and satellites have reported about 250 meteorites over the last ten years that have detonated at Earth's atmosphere. One of them had the explosive power equal to four or five times that of the atomic bomb which fell on Hiroshima. If a meteorite or asteroid with a diameter of five miles fell on the United States, it would cause its complete destruction. Asteroids capable of destroying a quarter of the world's population collide with Earth twice every million years. Smaller objects capable of destroying the population of a major city could hit once every two or three centuries. Impacts put the whole planet at risk. This is the only natural disaster that we may be able to do something about. A scientific meeting on the topic "The End of the World," attended by many countries was organized to discuss the steps to be taken in threats of meteorites, asteroids or comets approaching the Earth. It was suggested that the strongest long-range rocket in the world, the Russian Enagia, could be used to carry an American warhead for defence. This would be directed at an asteroid or meteorite with the intention of deflecting it from the Earth. Even if it is deflected, it could prove catastrophic because one of its fragments could hit the Earth.

The Qur'anic description of the universe's natural phenomena is further evidence of scientific miracles in the Qur'an. The description is accurate and forewarns us of possible dangers. The Prophet Muhammad could not read or write. The information could only have been produced from a scientific encyclopedia of the twentieth century.

Midway through the unfolding of the Qur'anic revelation, God completes His name *Al-Rahman* using the symbolic letters. He directs our attention to the signs in the Qur'an. The Arabic word for sign, *ayah,* also means "verse," but it has also been used to describe the miracles of Moses and other prophets. The signs or miracles in the Qur'an are given as science allusions. The universe was created in four dimensional space-time. It started as smoke and earth was separated from it. Before the Day of Judgement, the heavens and earth will change but will remain permanent after that. All of creation, including the Sun and the Moon, have a life span. Everything is balanced. God possesses the stores of all matter and He provides it in known quantities. The Moon's path passes through houses (constellations) and phases to teach us arithmetic and to provide us with a calendar. Together with the constellations, they guide us at night. Some of the stars are used as rockets. He described the *rawasi* or lithosphere as the extension and reduction of the earth and the mountains as anchored pegs. This is now known as the theory of plate tectonics. All plants are irrigated by the same water, but they are all different genetically. He uses wind for the transfer of pollens and He has all the stores of water. He describes how He created Man and made him His successor on earth. He recognizes his instincts that lead him to cruelty, lying, sexual temptation and murder, and his physiological signs of fear. His dreams are in code but can be interpreted by those with vision. He describes hysterical blindness and He is aware of what is in every woman's womb. He states that a child should not be separated from its mother before 30 months. He also confirms that He has no son or human parts.

The *surah*s beginning with the symbolic letters *alif lām mīm* contain 10.8 percent of the verses in the Qur'an. They have part of the main body of knowledge and advice for mankind's guidance. They contain references to the field of public health (including the diagnosis of pregnancy), descriptions of the water cycle, the physical universe, and the fourth dimension (time).

In many verses, the Qur'an discusses light, which seems to flow in a circuit. It starts with God, as the light of the universe, and it finishes with worthy people presenting their light to God on the Day of Judgment. One can understand this light circuit metaphorically on the basis that knowledge is light. On the other hand, the light circuit may be absolutely true light scientifically. Man is driven by energy. God put this energy in man's body in the form of spirit. Extra energy is given to people who are required to perform extra duties in the form of extra spirit or *rūḥ*. The earth is reduced when anybody dies. The human body does not disappear at death, rather, it remains on earth; only a small fraction will go to God to be preserved in a book. We can safely assume that this is our *rūḥ,* which would be of a known weight that can be measured by God.

God created man and made his body a complex of biochemical reactions leading to electric currents that move his muscles, his organs, and drive his thoughts and actions. We measure these electric currents of the heart by the ECG, the muscles by the EMG and the brain by the EEG. Even during our sleep the brain's thoughts and actions are evident from the EEG. Unfortunately, the EEG does not give very much detail of what is inside our brain, for example, it does not give the exact thought. The EEG is taken from outside the skull and simply records the sounds of the brain. This is similar to hundreds of people, men, women, and children, all speaking at once in a great hall that echoes. However, what is definitely proven is that our thoughts are recorded in electric impulses that appear as waves in an EEG. Excessive stimulation could prove pathological and cause disease, such as epilepsy. With this explanation one can understand that every thought or movement leads to an electric impulse or signal that is stored somewhere in the brain.

The light circuit in the Qur'an starts with God who is the light of the heavens and earth, even after the Sun disappears. The light from heaven is conveyed to Earth by a spirit from God's command that passed to Muhammad, thus giving him the revelation of the Qur'an. Such light was also sent to the Prophet Moses in his holy book, the Torah, and to other prophets.

The Qur'an possesses fundamental verses which are in fact *umm al-kitāb*. This is a heavenly book that is probably written in light and contains all the instructions and guidance that has been sent to man. On Earth it was transcribed and some of it is made into a readable holy book, the Qur'an, that can be studied by mankind.

God describes the Qur'an as High, Mighty, Wise, and enlightening. It is a manifest light that we should follow to bring us out of darkness into light.

Light propagates throughout the Qur'an in a circuit through all the verses of the Qur'an. It is channelled along God's name, *Al-Rahman*. The beginning and the middle of the name, recorded by the symbolic letters, at the beginning of *surah*s 50 through 66 (numbered cronologically). The name could only be completed by finding the *nūn* placed at the end of 3,123 verses which are spread throughout the Qur'an. It has been shown by radar chart that the letters *alif lām rā* act as two feeder systems in the form of two scalene, coinciding triangles that are directed to a central point where the *ḥā* *mīm* also accumulate and then spread out to the *nūn*s. If the radar graph is repeated using the cronological order of revelation, all the letters of the name *Al-Rahman* will concentrate in the central point except for a few scattered *nūn*s. The radar charts show scientifically that God's name *Al-Rahman* flows with the letters that make the name through and across a unique path in the Qur'an.

From His name, He overwhelms man with His mercy by giving him guidance. His guidance, contained in His messages to His prophets, followed the principles and advice provided by His names, thus fulfilling the objective of the Qur'an as a book of guidance and mercy. All this is given to man because God is equitable. Knowledge from the Qur'an is received in our brain as thoughts recorded by electric impulses. A special recording system in a concealed area in our brain, programmed by God, will store the information. The scoring system is in favor of man, only bad deeds but not bad thoughts are scored. Good thoughts and deeds are both scored, sometimes in doubles or multiples. People with high scores of good deeds will have higher light points that will shine around them on the Day of Judgment. It is quite possible that all the recorded information concerning man's thoughts or actions will go to God together with his soul when he dies. This would be kept in a book with God. Light points can only be gained on Earth. Those with high light points will be worthy of Paradise.

It seems that the symbolic letters initiate, propagate, and seal the path of the light circuit in the Qur'an. A very striking feature for any reader of the Qur'an, is the endless number of books mentioned in connection with the creation, the universe, the heavens and earth and what is in between, and the creatures including man.

There must be an infinite number of books, which cannot possibly be recorded, like our book, on paper. It must be recorded by some heavenly technique faster and more efficient than any known human method. It must have powers billions of times stronger than any human computer. The heavenly library must have an unbelievable power to collect, collate, file, store, and retrieve every minute detail of information present in each of these books.

The suggestion is that the Glorious Qur'an is provided with a light system initiated, propagated, and sealed by the symbolic letters. They sign the name of God, probably in light, across and through the Qur'an, going round in a circular motion to connect with the letter *nūn* at the end of 50.08 percent of the verses of the Qur'an. It is suggested that the symbolic letters act like computer markers, distinguishing the Qur'an in *umm al-kitāb* with God's name signed across and through it in glorious light. This would make the Qur'an stand out in *umm al-kitāb* among all other books.

BIBLIOGRAPHY

Abbas, Adel. M.A. *His Throne Was on Water.* amana publications, 1997.

Albritton, C.C. *Catastrophic Episodes in Earth's History.*

Bonthron, D.T. et al. "A Human Parthenogenetic Chimaera." *Natyre Genetics,* Vol II (October 1995): 164–169. Chapman & Hall, 1989.

Battan, L.J. *The Nature of Violent Storms.* Doubleday, 1961.

Binzel, R.P. et al. *Asteroids II.* University of Arizona Press, 1990.

Chorlton, Windsor. *Ice Ages.* Time Life, 1983.

Cunningham, C. *Introduction to Asteroids.* Willmann-Bell, 1987.

Hawkins, Gerald S. *The Physics and Astronomy of Meteors, Comets and Meteorites.* Magraw-Hill, 1964.

Hawkins, Stephen W. *A Brief History of Time.* Bantam Press, 1988.

Hutchinson Multimedia Encyclopedia 1995 (CD ROM). Attica and Helicon.

Imbrie, John & P. Katherine. *Ice Ages.* Harvard University Press, 1979.

McCall, G.J. *Meteorites and Their Origins.* Springer-Verlag, 1974.

Matthews, William H. *The Story of Glaciers and the Ice Age.* Harvey House, 1974.

Multimedia Encyclopedia (CD ROM). The Software Toolworks, 1992.

Osterbrock, D.E. & P. Raven. *Origins and Extinctions.* Yale University Press, 1988.

Readers Digest Book of Strange Stories and Amazing Facts. Readers Digest Association Limited, 1973.

Reihl, Herbert. *Climate and Weather in the Tropics.* Academic Press, 1979.

Sabbagh, K. *The Living Body.* Macdonald & Co., 1984.

Simpson, R.H. & H. Reil. *The Hurricane and Its Impact.* Academic Press, 1981.

Smith, Peter J. *Hutchinson Encyclopedia of the Earth.* Hutchinson, 1985.

Stark, David D. & William G. Bradley, Jr. *Magnetic Resonance Imaging.* 2nd ed. Vol. 1. Mosby Year Book Inc., 1991.

Ali, Yusuf. *The Holy Qur'an: Text Translation and Commentary.* amana publications, 1989.

Arberry, Arthur. *The Koran Interpreted.* Oxford University Press, 1983.

Maulvi, Muhammad Ali, *The Holy Qur'an.* Containing the Arabic Text with English Translation and Commentary. Ahmadiyya Anjuman-I-Ishaat-I-Islam, 1920.

Walton, John (editor). *Brain's Diseases of the Nervous System.* Oxford University Press, 1993.

Weems, John Edward. *The Tornado.* Doubleday, 1977.

Whipple, F.L. *The Mystery of the Comets.* Cambridge University Press, 1985.

Yeomans, Donald K. *Comets: A Chronological History of Observation, Science, Myth and Folklore.* Wiley, 1991.

APPENDIX

The Rhyme or the *Qāfiya*
at the End of the Verses of the Qur'an

SURAH 1 FATIHA: THE OPENING
Surah 1 Verse Number and Rhyming Letter

1	2	3	4	5	6	7			TOTAL
م	ن	م	ن	ن	م	ن			7

Surah 1 Proportional Distribution of Rhyming Letters

م	ن	TOTAL
3	4	7
42.8%	57.2%	100%

SURAH 2 AL BAQARA: THE HEIFER
Surah 2 Verse Number and Rhyming Letter

1	2-6	7	8-19	20	21-28	29	30-31	32	33-36
الم	ن	م	ن	ز	ن	م	ن	م	ن
37	38-36	49	50-53	54	55-103	104-105	106-107	108	109-110
م	ن	م	ن	م	ن	م	ر	ل	ر
111-113	114-115	116-118	119	120	121-124	125	126	127-129	130-136
ن	م	ن	م	ر	ن	د	ر	م	ن
137	138-141	142-143	144-147	148	149-157	158	159	160	161-162
م	ن	م	ن	ر	ن	م	ن	م	ن
163	164	165-166	167	168-172	173-174	175	176	177	178
م	ن	ب	ر	ن	م	ر	د	ن	م
179-180	181-182	183-191	192	193-195	196-197	198	199	200	201
ن	م	ن	م	ن	ب	ن	م	ق	ر
202	203	204	205-207	208	209	210	211-212	213	214
ب	ن	م	د	ن	م	ر	ب	م	ب
215	216-217	218	219	220	221-223	224-228	229-230	231	232

م	ن	م	ن	م	ن	م	ن	م	ن
233-234	235	236	237	238-239	240	241-243	244	245-246	247
ر	م	ن	ر	ن	م	ن	م	ن	م
248-252	253	254	255-256	257-258	259	260-261	262	263	264
ن	ر	ن	م	ن	ر	م	ن	م	ن
265	266	267	268	269	270-271	272	273	274-275	276
ر	ن	ر	م	ب	ر	ن	م	ن	م
277-281	282-283	284-285	286						
ن	م	ر	ن						

Surah 2 Proportional Distribution of Rhyming Letters

ن	م	ر	ل	د	ب	ق	TOTAL
193	54	21	1	7	9	1	286
67.8%	18.8%	7.2%	0.34%	2.44%	3.14%	0.34%	100%

SURAH 3 AL-'IMRAN: THE FAMILY OF 'IMRAN
Surah 3 Verse Number and Rhyming Letter

1	2	3	4	5	6	7-8	9	10	11
ر	ن	ل	م	ب	ر	ن	م	ن	م
12	13	14	15	16-17	18	19	20	21	22-25
د	ر	ب	د	ر	م	ب	د	م	ن
26	27	28-29	30	31	32-33	34-36	37	38	39
ر	ب	ر	د	م	ن	م	ب	ء	ن
40	41	42-47	48	49-50	51	52-57	58	59-61	62
ء	ر	ن	ل	ن	م	ن	م	ن	م
63-72	73-74	75-76	77	78-88	89	90-91	92	93-100	101
ن	م	ن	م	ن	م	ن	م	ن	م
102-104	105	106-108	109	110-118	119	120	121	122-125	126
ن	م	ن	ر	ن	ر	ط	م	ن	م

Science Miracles: No Stick or Snakes

127-128	129	130-153	154	155	156	157-161	162	163-164	165
ن	م	ن	ر	م	ر	ن	ر	ن	ر
166-171	172	173	174	175	176-177	178	179	180	181
ن	م	ل	م	ن	م	ن	م	ر	ق
182	183	184-186	187	188	189	190	191-193	194	195
د	ن	ر	ن	م	ر	ب	ر	د	ب
196-197	198	199	200						
د	ر	ب	ن						

Surah 3 Proportional Distribution of the Rhyming Letters

ن	م	ب	ر	د	ل	ء	ط	ق	Total
121	30	9	23	9	3	3	1	1	200
60.5%	15%	4.5%	11.5%	4.5%	1.5%	1.5%	0.5%	0.5%	100%

SURAH 4 NISAA:THE WOMEN
Surah 4 Verse Number and Rhyming Letter

1-11	12-13	14	15-24	25-26	27-43	44	45-175	176	
ا	م	ن	ا	م	ا	ل	ا	م	

Surah 4 Proportional Distribution of the Rhyming Letters

م	ا	ل	TOTAL
5	170	1	176
2.81%	96.6%	0.56%	100%

SURAH 5 MA'IDAH:THE TABLE
Surah 5 Verse Number and Rhyming Letter

1	2	3	4	5-6	7	8	9-10	11	12
د	ب	م	ب	ن	ر	ن	م	ن	ل
13-15	16	17-19	20-32	33-34	35	36-39	40	41	42-53
ن	م	ر	ن	م	ن	م	ر	م	ن
54	55-59	60	61-64	65	66-71	72	73-74	75	76
م	ن	ل	ن	م	ن	ر	م	ن	م

Surah 5 (continued) Verse Number and Rhyming Letter

77	78-85	86	87-93	94-95	96	97-98	99-100	101	102-108
ل	ن	م	ن	م	ن	م	ن	م	ن
109	110-115	116	117	118-119	120				
ب	ن	ب	د	م	ر				

Surah 5 Proportional Distribution of Rhyming Letters

ن	م	د	ب	ل	ر	TOTAL
80	24	2	4	3	7	120
66.66%	20%	1.66%	3.3%	2.5%	5.8%	100%

SURAH 6 AN'AM: CATTLE
Surah 6 Verse Number and Rhyming Letter

1-12	13	14	15	16	17-18	19-38	39	40-53	54
ن	م	ن	م	ن	ر	ن	م	ن	م
55-65	66	67-72	73	74-82	83	84-86	87	88-95	96
ن	ل	ن	ر	ن	م	ن	م	ن	م
97-100	101	102	103	104	105-106	107	108-114	115	116-127
ن	م	ل	ر	ظ	ن	ل	ن	م	ن
128	129-138	139	140-144	145	146-164	165			
م	ن	م	ن	م	ن	م			

Surah 6 Proportional Distribution of Rhyming Letters

ن	م	ر	ل	ظ	TOTAL
144	13	4	3	1	165
87.2%	7.8%	2.4%	1.8%	0.6%	100%

SURAH 7 A'RAF: THE HEIGHTS
Verse Number and Rhyming Letter

1	2-15	16	17-58	59	60-72	73	74-104	105	106-108
المص	ن	م	ن	م	ن	م	ن	ل	ن
109	110-111	112	113-115	116	117-133	134	135-140	141	142-152
م	ن	م	ن	م	ن	ل	ن	م	ن

153	154-166	167	168-199	200	201-206			
م	ن	م	ن	م	ن			

Surah 7 Proportional Distribution of Rhyming Letters

ن	م	ل	ص	TOTAL
193	10	2	1	206
93.68%	4.85%	0.97%	0.48%	100%

SURAH 8 ANFAL:SPOILS of WAR
Surah 8 Verse Number and Rhyming Letter

1-3	4	5-9	10-11	12	13	14-16	17	18-24	25
ن	م	ن	م	ن	ب	ر	م	ن	ب

26-27	28-29	30-31	32	33-38	39-41	42	43-44	45-46	47
ن	م	ن	م	ن	ر	م	ر	ن	ط

48	49	50	51	52	53	54-60	61	62	63
ب	م	ق	د	ب	م	ن	م	ن	م

64-66	67-71	72-73	74-75						
ن	م	ر	م						

Surah 8 Proportional Distribution of Rhyming Letters

ن	م	ب	ر	د	ط	ق	TOTAL
39	19	4	10	1	1	1	75
52%	25.3%	5.3%	13.35%	1.3%	1.3%	1.3%	100%

SURAH 9 TAUBAH:REPENTANCE
Surah 9 Verse Number and Rhyming Letter

1-2	3	4	5	6-14	15	16-20	21-22	23-26	27-28
ن	م	ن	م	ن	م	ن	م	ن	م

29-33	34	35-37	38	39	40	41-59	60-61	62	63
ن	م	ن	ل	ر	م	ن	م	ن	م

64-67	68	69-70	71-72	73-74	75-77	78	79	80-88	89-91
ن	م	ن	م	ر	ن	ب	م	ن	م

92-96	97-104	105	106	107-109	110-111	112	113-115	116	117-118
ن	م	ن	م	ن	م	ن	م	د	م
119-127	128-129								
ن	م								

Surah 9 Proportional Distribution of Rhyming Letters

ن	م	ر	ل	ب	TOTAL
86	37	4	1	1	129
66.66%	28.7%	3.1%	0.77%	0.77%	100%

SURAH 10 YUNUS:JONAH
Surah 10 Verse Numbers and Rhyming Letter

1	2-8	9	10-14	15	16-24	25	26-63	64-65	66-78
م	ن	م	ن	م	ن	م	ن	م	ن
79	80-87	88	89	90-96	97	98-106	107	108	109
م	ن	م	ن	ن	م	ن	م	ل	ن

Surah 10 Proportional Distribution of Rhyming Letters

ن	م	ل	TOTAL
98	10	1	109
89.9%	9.1%	0.9%	100%

SURAH 11 HUD:HOOD
Surah 11 Verse Number and Rhyming Letter

1-5	6-8	9-11	12	13-25	26	27-38	39	40	41
ر	ن	ر	ل	ن	م	ن	م	ل	م
42-47	48	49-55	56	57-58	59-60	61-62	63	64-65	66
ن	م	ن	م	ظ	د	ب	ر	ب	ز
67	68	69	70	71-72	73	74	75	76	77
ن	د	ذ	ط	ب	د	ط	ب	د	ب
78-80	81	82-83	84	85	86	87	88	89-90	91
د	ب	د	ط	ن	ظ	د	ب	د	ذ

Science Miracles: No Stick or Snakes

92	93	94	95	96	97-100	101	102-105	106	107
ط	ب	ن	د	ن	د	ب	د	ق	د
108	109	110	111-112	113-123					
ذ	ص	ب	ر	ن					

Surah 11 Proportional Distribution of the Rhyming Letters

ن	ر	ذ	د	ب	ل	ز	ط	ظ	ق	ص	م	TOTAL
56	11	3	23	11	2	1	4	3	1	1	5	123
45.5%	8.9%	2.4%	18.6%	10.5%	1.6%	0.8%	3.2%	2.4%	0.8%	0.8%	4%	100%

SURAH 12 YUSUF:JOSEPH
Surah 12 Verse Number and Rhyming Letter

1-5	6	7-24	25	26-27	28	29-30	31	32-33	34
ن	م	ن	م	ن	م	ن	م	ن	م
35-38	39	40-49	50	51-52	53	54	55	56-64	65
ن	ر	ن	م	ن	م	ن	م	ن	د
66	67-71	72	73-75	76	77-82	83-84	85-94	95	96-97
ل	ن	م	ن	م	ن	م	ن	م	ن
98	99	100	101-111						
م	ن	م	ن						

Surah 12 Proportional Distribution of the Rhyming Letters

ن	م	ل	د	ر	TOTAL
93	15	1	1	1	111
83.8%	13.5%	0.9%	0.9%	0.9%	100%

SURAH 13 R'AD:THUNDER
Verse Number and Rhyming Letter

1-5	6	7	8	9	10	11-15	16	17	18
ن	ب	د	ر	ل	ر	ل	ر	ل	د
19	20	21	22	23	24-25	26	27-30	31	32
ب	ق	ب	ر	ب	ر	ع	ب	ر	ب

33	34	35	36	37	38-41	42	43		
د	قَ	ر	ب	قَ	ب	ر	ب		

Surah 13 Proportional Distribution of the Rhyming Letters

ن	ب	ع	د	ر	ل	قَ	TOTAL
5	15	1	4	8	7	3	43
11.6%	34.9%	2.3%	9.3%	18.6%	16.3%	7%	100%

SURAH 14 IBRAHIM:ABRAHAM
Surah 14 Verse Number and Rhyming Letter

1-3	4	5	6	7-8	9	10-13	14-16	17	18-19
د	م	ر	م	د	ب	ن	د	ظ	د
20	21	22-23	24	25	26	27	28-30	31	32-34
ز	ص	م	ء	ن	ر	ء	ر	ل	ر
35-36	37	38-40	41	42	43	44-46	47	48	49
م	ن	ء	ب	ر	ء	ل	م	ر	د
50	51-52								
ر	ب								

Surah 14 Proportional Distribution of Rhyming Letters

ن	م	ر	د	ب	ظ	ز	ص	ء	ل	TOTAL
6	7	14	10	4	1	1	1	4	4	52
11.5%	13.5%	27%	19.2%	7.7%	1.9%	1.9%	1.9%	7.6%	7.6%	100%

SURAH 15 HIJR:ROCKY TRACT
Surah 15 Verse Number and Rhyming Letter

1-3	4	5-16	17	18-20	21	22-24	25	26	27
ن	م	ن	م	ن	م	ن	م	ن	م
28-33	34	35-37	38	39-40	41	42-43	44	45-48	49-51
ن	م	ن	م	ن	م	ن	م	ن	م

Science Miracles: No Stick or Snakes

52	53	54-73	74	75	76	77-84	85	86-87	88-99
ن	م	ن	ل	ن	م	ن	ل	م	ن

* pronounced like Alif ا

Surah 15 Proportional Distribution of Rhyming Letters

ن	م	ل	TOTAL
81	16	2	99
81.8%	16.2%	2%	100%

SURAH 16 NAHML: THE BEE
Surah 16 Verse Number and Rhyming Letter

1-6	7	8-17	18	19-46	47	48-57	58	59	60
ن	م	ن	م	ن	م	ن	م	ن	م

61-62	63	64-69	70	71-75	76	77	78-93	94	95-97
ن	م	ن	ر	ن	م	ر	ن	م	ن

98	99-103	104	105	106	107-109	110	111-114	115	116
م	ن	م	ن	م	ن	م	ن	م	ن

117	118	119	120	121	122-128				
م	ن	م	ن	م	ن				

Proportional Distribution of Rhyming Letters

ن	م	ر	TOTAL
110	16	2	128
85.9%	12.5%	1.6%	100%

SURAH 17 AL-ISRA: THE NIGHT JOURNEY
Surah 17 Verse Number and Rhyming Letter

1	2-111								
ر	ا								

Surah 17 Proportional Distribution of Rhyming Letters

ر	ا	TOTAL
1	110	111
0.9%	99.1%	100%

SURAH 18 KAHF: THE CAVE
Surah 18 Verse Number and Rhyming Letter

1-12	13	14-110						
ا	*ي	ا						

*pronounced like Alif ا .

Surah 18 Proportional Distribution of Rhyming Letters

ا	*ي	TOTAL
109	1	110
99.1%	0.9%	100%

SURAH 19 MARYAM: MARY
Surah 19 Verse Numbers and Rhyming Letter

1	2-33	34-35	36-37	38-40	41-98				
ص	ا	ن	م	ن	ا				

Surah 19 Proportional Distribution of Rhyming Letters

ص	ن	ا	م	TOTAL
1	5	90	2	98
1%	5.1%	91.8%	2%	100%

SURAH 20 TA-HA
Surah 20 Verse Number and Rhyming Letter

1	2-24	25-32	33-35	36-38	39	40	41-42	43-71	72
ه	*ي	ي	ا	*ي	ي	*ي	ي	*ي	ا
73-77	78	79-84	85-88	89	90-91	92	93-96	97-115	116-124
*ي	م	*ي	ي	ا	*ي	ا	ي	ا	*ي
125	126-135								
ا	*ي								

* pronounced like Alif ا

Surah 20 Proportional Distribution of the Rhyming Letters

ي	*ي	م	ا	ه	TOTAL
19	88	1	26	1	135
14%	65.2%	0.7%	19.3%	0.7%	100%

SURAH 21 ANBIYA:THE PROPHETS
Surah 21 Verse Number and Rhyming Letter

1-3	4	5-59	60	61	62	63-65	66	67-68	69
ن	م	ن	م	ن	م	ن	م	ن	م
70-75	76	77-112							
ن	م	ن							

Surah 21 Proportional Distribution of Rhyming Letters

ن	م	TOTAL
106	6	112
94.6%	5.4%	100%

SURAH 22 HAJJ:PILGRIMAGE
Surah 22 Verse Number and Rhyming Letter

1	2-3	4	5	6-8	9	10	11	12	13
م	د	ر	ج	ر	قَ	د	ن	د	ر
14	15	16-17	18	19	20-21	22	23	24	25
د	ظ	د	ء	م	د	قَ	ر	د	م
26	27	28	29	30	31	32	33	34-37	38-39
د	قَ	ر	قَ	ر	قَ	ب	قَ	ن	ر
40	41	42	43	44	45	46	47	48	49
ز	ر	د	ط	ر	د	ر	ن	ر	ن
50-52	53	54-56	57-58	59	60-63	64	65	66	67
م	د	م	ن	م	ر	د	م	ر	م
68-69	70-72	73	74	75-76	77	78			
ن	ر	ب	ز	ر	ن	ر			

Surah 22 Proportional Distribution of Rhyming Letters

ن	م	د	ر	ج	قَ	ب	ظ	ء	ط	ز	TOTAL
12	12	15	25	1	6	2	1	1	1	2	78
15.4%	15.4%	19.2%	32.1%	1.3%	7.7%	2.6%	1.3%	1.3%	1.3%	2.6%	100%

SURAH 23 MU'MINUN:THE BELIEVERS
Surah 23 Verse Number and Rhyming Letter

1-50	51	52-72	73	74-85	86	87-115	116	117-118	
ن	م	ن	م	ن	م	ن	م	ن	

Surah 23 Proportional Distribution of Rhyming Letters

ن	م	TOTAL
114	4	118
96.6%	3.4%	100%

SURAH 24 NUR:LIGHT
Surah 24 Verse Number and Rhyming Letter

1-4	5	6-9	10-11	12-13	14-16	17	18	19	20-23
ن	م	ن	م	ن	م	ن	م	ن	م
24-25	26	27	28	29-31	32-33	34	35	36	37
ن	م	ن	م	ن	م	ن	م	ل	ر
38-39	40	41	42-45	46	47-56	57	58-60	61	62-64
ب	ر	ن	ر	م	ن	ر	م	ن	م

Surah 24 Proportional Distribution of Rhyming Letters

ن	م	ل	ر	ب	TOTAL
31	23	1	7	2	64
48.4%	35.9%	1.6%	10.9%	3.1%	100%

SURAH 25 FURQAN:THE CRITERION
Surah 25 Verse Number and Rhyming Letter

1-16	17	18-77							
ا	ل	ا							

Surah 25 Proportional Distribution of Rhyming Letters

ا	ل	TOTAL
76	1	77
98.7%	1.3%	100%

SURAH 26 SHU'ARA: THE POETS
Surah 26 Verse Number and Rhyming Letter

1	2-6	7	8	9	10-16	17	18-21	22	23-33
م	ن	م	ن	م	ن	ل	ن	ل	ن
34	35-36	37-38	39-57	58	59	60-62	63	64-67	68-69
م	ن	م	ن	م	ل	ن	م	ن	م
70-84	85	86-88	89	90-100	101	102-103	104	105-121	122
ن	م	ن	م	ن	م	ن	م	ن	م
123-134	135	136-139	140	141-147	148	149-154	155-156	157-158	159
ن	م	ن	م	ن	م	ن	م	ن	م
160-174	175	176-181	182	183-188	189	190	191	192-196	197
ن	م	ن	م	ن	م	ن	م	ن	ل
198-200	201	202-216	217-218	219	220	221	222	223-227	
ن	م	ن	م	ن	م	ن	م	ن	

Surah 26 Proportional Distribution of Rhyming Letters

ن	م	ل	TOTAL
192	31	4	227
84.6%	13.7%	1.8%	100%

SURAH 27 NAML: THE ANTS
Surah 27 Verse Number and Rhyming Letter

1-5	6	7-8	9	10	11	12-22	23	24-25	26
ن	م	ن	م	ن	م	ن	م	ن	م
27-28	29-30	31-39	40	41-77	78	79-93			
ن	م	ن	م	ن	م	ن			

Surah 27 Proportional Distribution of Rhyming Letters

ن	م	TOTAL
84	9	93
90.3%	9.67%	100%

SURAH 28 QASAS: THE NARRATION
Surah 28 Verse Number and Rhyming Letter

1	2-15	16	17-21	22	23-24	25-27	28	29-78	79
م	ن	م	ن	ل	ر	ن	ل	ن	م
80-88									
ن									

Surah 28 Proportional Distribution of Rhyming Letters

ن	م	ل	ر	TOTAL
81	3	2	2	88
92%	3.4%	2.3%	2.3%	100%

SURAH 29 ANKABUT: THE SPIDER
Surah 29 Verse Number and Rhyming Letter

1	2-4	5	6-18	19-20	21	22	23	24-25	26
م	ن	م	ن	ر	ن	ر	م	ن	م
27-41	42	43-59	60	61	62	63-69			
ن	م	ن	م	ن	م	ن			

Surah 29 Proportional Distribution of Rhyming Letters

ن	م	ر	TOTAL
59	7	3	69
85.5%	10.1%	4.3%	100%

SURAH 30 RUM: THE ROMANS
Surah 30 Verse Number and Rhyming Letter

1	2	3-4	5	6-26	27	28-49	50	51-52	54
م	م	ن	م	ن	م	ن	ر	ن	ر
55-60									
ن									

Surah 30 Proportional Distribution of Rhyming Letters

ن	م	د	TOTAL
54	4	2	60
90%	6.66%	3.33%	100%

SURAH 31 LUQMAN:LOCKMAN
Surah 31 Verse Number and Rhyming Letter

1	2	3-6	7-10	11	12	13	14	15	16-23
م	م	ن	م	ن	د	م	ر	ن	ر

24	25	26	27	28-34					
ط	ن	د	م	ر					

Surah 31 Proportional Distribution of Rhyming Letters

ن	م	د	ر	ط	TOTAL
7	8	2	16	1	34
20.6%	23.5%	5.9%	47.1%	2.9%	100%

SURAH 32 SAJDAH:ADORATION
Surah 32 Verse Number and Rhyming Letter

1	2-5	6	7-22	23	24-30				
م	ن	م	ن	ل	ن				

Surah 32 Proportional Distribution of Rhyming Letters

ن	م	ل	TOTAL
27	2	1	30
90%	6.66%	3.33%	100%

SURAH 33 AHZAB:THE CONFEDERATES
Surah 33 Verse Number and Rhyming Letter

1-3	4	5-73							
ا	ل	ا							

Surah 33 Proportional Distribution of Rhyming Letters

ا	ل	TOTAL
72	1	73
98.6%	1.4%	100%

SURAH 34 SABA: THE CITY OF SABA

Surah 34 Verse Number and Rhyming Letter

1-2	3	4-5	6-8	9	10	11-13	14	15	16
ر	ن	م	د	ب	د	ر	ن	ر	ل
17	18	19	20	21	22-23	24-25	26-27	28-43	44-45
ر	ن	ر	ن	ظ	ر	ن	م	ن	ر
46-47	48	49	50-51	52-53	54				
د	ب	د	ب	د	ب				

Surah 34 Proportional Distribution of Rhyming Letters

ن	م	ر	ل	د	ب	ظ	TOTAL
22	4	12	1	9	5	1	54
40.7%	7.4%	22.22%	1.9%	16.66%	9.3%	1.19%	100%

SURAH 35 FATIR: THE ORIGINATOR OF CREATION

Surah 35 Verse Number and Rhyming Letter

1	2	3	4-7	8	9-11	12	13-14	15-16	17
ر	م	ن	ر	ن	ر	ن	ر	د	ز
18-26	27	28-34	35	36-38	39-45				
ر	د	ر	ب	ر	ا				

Surah 35 Proportional Distribution of Rhyming Letters

ن	ر	م	د	ا	ب	ز	TOTAL
3	29	1	3	7	1	1	45
6.66%	64.4%	2.2%	6.66%	15.5%	2.2%	2.2%	100%

SURAH 36   YASIN

Surah 36 Verse Number and Rhyming Letter

1	2	3	4-5	6-10	11	12-17	18	19-37	38
س	م	ن	م	ن	م	ن	م	ن	م
39-57	58	59-60	61	62-77	78-79	80	81	82-83	
ن	م	ن	م	ن	م	ن	م	ن	

Surah 36 Proportional Distribution of Rhyming Letters

ن	م	س	TOTAL
71	11	1	83
85.5%	13.3%	1.2%	100%

SURAH 37 SAFFAT: THOSE RANGED IN RANKS

Surah 37 Verse Number and Rhyming Letter

1-3	4	5	6	7	8-11	12-22	23	24-37	38
ا	د	ق	ب	د	ب	ن	م	ن	م
39-40	41	42	43	44-54	55	56-59	60	61	62
ن	م	ن	م	ن	م	ن	م	ن	م
63	64	65-66	67-68	69-75	76	77-82	83-84	85-87	88-89
ن	م	ن	م	ن	م	ن	م	ن	م
90-96	97	98-100	101	102-103	104	105-106	107	108	109
ن	م	ن	م	ن	م	ن	م	ن	م
110-114	115	116-117	118	119-141	142	143-144	145	146-162	163-164
ن	م	ن	م	ن	م	ن	م	ن	م
165-182									
ن									

Surah 37 Proportional Distribution of Rhyming Letters

ن	م	ا	د	ب	ق	TOTAL
145	26	3	2	5	1	182
79.7%	14.3%	1.6%	1.1%	2.7%	0.54%	100%

SURAH 38 SAD

Surah 38 Verse Number and Rhyming Letter

1	2	3	4-5	6	7	8-11	12	13-14	15
ر	ق	ص	ب	د	ق	ب	د	ب	ق
16-17	18	19-21	22	23-26	27-28	29-30	31	32	33
ب	ق	ب	ط	ب	ر	ب	د	ب	ق
34-36	37	38	39-44	45-48	49-53	54	55	56	57
ب	ص	د	ب	ر	ب	د	ب	د	ق
58	59-66	67	68-76	77	78-80	81	82-83	84	85-88
ج	ر	م	ن	م	ن	م	ن	ل	ن

Surah 38 Proportional Distribution of Rhyming Letters

ن	ص	ر	ق	ب	د	ط	ج	م	ل	TOTAL
18	2	15	6	35	6	1	1	3	1	88
20.5	2.3%	17%	6.8%	39.8	6.81	1.1%	1.1%	3.4%	1.1%	100

SURAH 39 ZUMAR: THE CROWDS
Surah 39 Verse Numbers and Rhyming Letter

1	2	3-5	6	7-8	9-10	11-12	13	14	15-16
م	ن	ر	ن	ر	ب	ن	م	ي	ن
17	18	19	20	21	22	23	24-35	36	37
د	ب	ر	د	ب	ن	د	ن	د	م
38-39	40	41	42-52	53	54-61	62	63-75		
ن	م	ل	ن	م	ن	ل	ن		

Surah 39 Proportional Distribution of Rhyming Letters

ن	م	ر	ب	ي	ل	د	TOTAL
53	5	6	4	1	2	4	75
70.7%	6.66%	8%	5.3%	1.3%	2.66%	5.3%	100%

SURAH 40 GHAFIR: THE BELIEVERS
Surah 40 Verse Number and Rhyming Letter

1	2	3	4	5	6	7-9	10	11	12
م	م	ر	د	ب	ر	م	ن	ل	ر
13	14	15	16	17	18	19-20	21	22	23
ب	ن	ق	ر	ب	ع	ر	ق	ب	ن
24	25	26	27-28	29	30	31-33	34	35	36-37
ب	ل	د	ب	د	ب	د	ب	ر	ب
38	39	40	41-43	44	45-46	47	48	49	50
د	ر	ب	ر	د	ب	ر	د	ب	ل
51	52	53-54	55-56	57-85					
د	ر	ب	ر	ن					

Surah 40 Proportional Distribution of Rhyming Letters

ن	م	ب	ر	د	ل	ق	ع	TOTAL
32	5	17	15	10	3	2	1	85
37.6%	5.9%	20%	17.6%	11.76%	3.5%	2.4%	1.1%	100%

SURAH 41 FUSSILAT:
Surah 41 Verse Number and Rhyming Letter

1	2	3-11	12	13	14-31	32	33	34-36	37-38
م	م	ن	م	د	ن	م	ن	م	ن

39-40	41	42	43	44	45	46-47	48	49	50
ر	ز	د	م	د	ب	د	ص	ط	ظ

51	52-53	54							
ض	د	ط							

Surah 41 Proportional Distribution of Rhyming Letters

ن	م	د	ر	ز	ب	ص	ض	ط	ظ	TOTAL
30	8	7	2	1	1	1	1	2	1	54
55.5%	14.8%	13%	3.7%	1.85%	1.85%	1.85%	1.85%	3.7%	1.85%	100%

SURAH 42 SHURA:CONSULTATION
Surah 42 Verse Number and Rhyming Letter

1	2	3-5	6	7-9	10	11	12	13-14	15
م	قَ	م	ل	ر	ب	ر	م	ب	ر

16	17	18	19	20	21	22-24	25	26	27
د	ب	د	ز	ب	م	ر	ن	د	ر

28	29-31	32	33-34	35	36-40	41	42	43	44
د	ر	م	ر	ص	ن	ل	م	ر	ل

45	46	47-50	51-52	53					
م	ل	ر	م	ر					

Surah 42 Proportional Distribution of Rhyming Letters

ن	م	قَ	ل	ر	د	ز	ص	ب	TOTAL
6	11	1	4	20	4	1	1	5	53
11.3%	20.8%	1.88%	7.5%	37.7%	7.5%	1.88%	1.88%	9.4%	100%

SURAH 43 ZUKHAF:GOLD ADORNMENTS
Surah 43 Verse Number and Rhyming Letter

1	2-3	4	5-8	9	10-16	17	18-30	31	32-42
م	ن	م	ن	م	ن	م	ن	م	ن

43	44-58	59	60	61	62-63	64-65	66-83	84	85-89
م	ن	ل	ن	م	ن	م	ن	م	ن

Surah 43 Proportional Distribution of Rhyming Letters

ن	م	ل	TOTAL
78	10	1	89
87.6%	11.2%	1.1%	100%

SURAH 44 DUKHAN:SMOKE
Surah 44 Verse Number and Rhyming Letter

1	2-3	4	5	6	7-10	11	12-16	17	18-25
م	ن	م	ن	م	ن	م	ن	م	ن

26	27-41	42-44	45	46-49	50-55	56-57	58-59		
م	ن	م	ن	م	ن	م	ن		

Surah 44 Proportional Distribution of Rhyming Letters

ن	م	TOTAL
44	15	59
74.6%	25.4%	100%

SURAH 45 JATHIYA:KNEELING
Surah 45 Verse Number and Rhyming Letter

1	2	3-6	7-8	9	10-11	12-36	37		
م	م	ن	م	ن	م	ن	م		

Surah 45 Proportional Distribution of Rhyming Letters

ن	م	Total
30	7	37
81.1%	18.9%	100%

SURAH 46 AHQAF: WINDING SAND-TRACTS
Surah 46 Verse Number and Rhyming Letter

1	2	3-7	8	9-10	11	12-20	21	22-23	24
م	م	ن	م	ن	م	ن	م	ن	م

25-29	30-31	32	33	34-35					
ن	م	ن	ر	ن					

Surah 46 Proportional Distribution of Rhyming Letter

ن	م	ر	TOTAL
26	8	1	35
74.3%	22.9%	2.9%	100%

SURAH 47 MUHAMMAD
Surah 47 Verse Number and Rhyming Letter

1-9	10	11-23	24	25-38					
م	ا	م	ا	م					

Surah 47 Proportional Distribution of Rhyming Letters

م	ا	TOTAL
36	2	38
94.7%	5.3%	100%

SURAH 48 F'AT-H: VICTORY
Surah 48 Verse Number and Rhyming Letter

1-29									
ا									

Surah 48 Proportional Distribution of Rhyming Letters

ا	TOTAL
29	29
100%	100%

SURAH 49 HUJURAT:INNER APARTMENTS
Surah 49 Verse Number and Rhyming Letter

1	2	3	4	5	6-7	8	9-11	12	13
م	ن	م	ن	م	ن	م	ن	م	ر

14	15	16	17-18						
م	ن	م	ن						

Surah 49 Proportional Distribution of Rhyming Letters

ن	م	ر	TOTAL
10	7	1	18
55.5%	38.9%	5.6%	100%

SURAH 50 QAF
Surah 50 Verse Number and Rhyming Letter

1	2	3	4	5-7	8	9-10	11	12	13
د	ب	د	ط	ج	ب	د	ج	د	ط

14-24	25	26-31	32	33	34-35	36	37	38-39	40
د	ب	د	ظ	ب	د	ص	د	ب	د

41	42	43-44	45						
ب	ج	ر	د						

Surah 50 Proportional Distribution of Rhyming Letters

د	ب	ظ	ج	ص	ط	ر	TOTAL
27	7	1	5	1	2	2	45
60%	15.6%	2.2%	11.11%	2.2%	4.4%	4.4%	100%

SURAH 51 ZARIYAT:THE WINDS THAT SCATTER
Surah 51 Verse Number and Rhyming Letter

1-4	5	6	7	8	9	10-18	19	20-27	28-30
ا	ق	ع	ك	ف	ك	ن	م	ن	م

31-36	37	38-39	40-42	43-53	54	55-60			
ن	م	ن	م	ن	م	ن			

Surah 51 proportional Distribution of Rhyming Letters

ن	م	ا	ق	ك	ع	ف	TOTAL
42	9	4	1	2	1	1	60
70%	15%	6.66%	1.66%	3.33%	1.66%	1.66%	100%

SURAH 52 TUR: THE MOUNT
Surah 52 Verse Number and Rhyming Letter

1-4	5	6	7-8	9-10	11-12	13	14-16	17-18	19-22
ر	ع	ر	ع	ا	ن	ا	ن	م	ن

23	24-26	27-28	29-43	44	45-47	48 49			
م	ن	م	ن	م	ن	م			

Surah 52 Proportional Distribution of Rhyming Letters

ن	م	ع	ر	ا	TOTAL
30	8	3	5	3	49
61.2%	16.3%	6.1%	10.2%	6.1%	100%

SURAH 53 NAJM: THE STAR
Surah 53 Verse Number and Rhyming Letter

1-27	28-29	30-43	44	45-56	57-58	59-61	62		
*ي	ا	*ي	ا	*ي	ة	ن	ا		

* pronounced like Alif ا

Surah 53 Proportional Distribution of Rhyming Letters

ن	*ي	ا	ة	TOTAL
3	53	4	2	62
4.8%	85.5%	6.5%	3.2%	100%

* pronounced like Alif ا

SURAH 54 QUAMAR:MOON
Surah 54 Verse Number and Rhyming Letter

1-55								
ر								

Surah 54 Proportional Distribution of Rhyming Letter

ر	TOTAL
55	55
100%	100%

SURAH 55 RAHMAN
Surah 55 Verse Number and Rhyming Letter

1-9	10-11	12-13	14-15	16-23	24	25-26	27	28-40	41
ن	م	ن	ر	ن	م	ن	م	ن	م

42-71	72	73-77	78
ن	م	ن	م

Surah 55 Proportional Distribution of Rhyming Letters

ن	م	ر	TOTAL
68	8	2	78
87.2%	10.3%	2.6%	100%

SURAH 56 WAQI'A:THE ENEVETABLE EVENT
Surah 56 Verse Number and Rhyming Letter

1-3	4-6	7-9	10-11	12	13-14	15	16-24	25-26	27
ة	ا	ة	ن	م	ن	ة	ن	ا	ن

28-30	31	32-34	35	36-37	38-40	41	42-44	45	46
د	ب	ة	ء	ا	ن	ل	م	ن	م

47-49	50	51	52	53	54-55	56-73	74-77	78-82	83
ن	م	ن	م	ن	م	ن	م	ن	م

84-88	89	90-92	93-94	95	96
ن	م	ن	م	ن	م

Surah 56 Proportional Distribution of Rhyming Letters

ن	م	ة	ا	د	ب	ء	ل	TOTAL
55	18	10	7	3	1	1	1	96
57.3%	18.6%	10.4%	7.3%	3.1%	1%	1%	1%	100%

SURAH 57 HADEED:IRON
Surah 57 Verse Number and Rhyming Letter

1	2	3	4-7	8	9	10	11-12	13	14-15
م	ر	م	ر	ن	م	ر	م	ب	ر
16-17	18-19	20	21	22-23	24	25	26-27	28-29	
ن	م	ر	م	ر	د	ز	ن	م	

Surah 57 Proportional Distribution of Rhyming Letters

ن	م	ر	ز	د	TOTAL
5	10	12	1	1	29
17.2%	34.5%	41.4%	3.4%	3.4%	100%

SURAH 58 MUJADILA:THE WOMAN WHO PLEADS
Surah 58 Verse Number and Rhyming Letter

1-3	4	5	6	7	8	9-10	11	12	13-20
ر	م	ن	د	م	ر	ن	ر	م	ن
21	22								
ز	ن								

Surah 58 Proportional Distribution of Rhyming Letters

ن	م	ر	ز	د	TOTAL
12	3	5	1	1	22
54.5%	13.6%	22.7%	4.5%	4.5%	100%

SURAH 59 HASHR:THE GATHERING
Surah 59 Verse Number and Rhyming Letter

1	2-3	4	5	6	7	8-9	10	11-14	15
م	ر	ب	ن	ر	ب	ن	م	ن	م

16-21	22	23	24						
ن	م	ن	م						

Surah 59 Proportional Distribution of Rhyming Letters

ن	م	ر	ب	TOTAL
14	5	3	2	24
58.3%	20.8%	12.5%	8.3%	100%

SURAH 60 MUMTAHANA: THE WOMAN TO BE EXAMINED

Surah 60 Verse Number and Rhyming Letter

1	2	3-4	5	6	7	8-9	10	11	12
ل	ن	ر	م	د	م	ن	م	ن	م
13									
ر									

Surah 60 Proportional Distribution of Rhyming Letters

ن	م	ل	ر	د	TOTAL
4	4	1	3	1	13
30.8%	30.8%	7.7%	23.1%	7.7%	100%

SURAH 61 SAFF: THE BATTLE ARRAY

Surah 61 Verse Number and Rhyming Letter

1	2-3	4	5-9	10	11	12	13-14		
م	ن	ص	ن	م	ن	م	ن		

Surah 61 Proportional Distribution of Rhyming Letters

ن	م	ص	TOTAL
10	3	1	14
71.4%	21.4%	7.1%	100%

SURAH 62 JUMU'AH: THE ASSEMBLY (FRIDAY)

Surah 62 Verse Number and Rhyming Letter

1	2	3-4	5-11						
م	ن	م	ن						

Surah 62 Proportional Distribution of Rhyming Letters

ن	م	TOTAL
8	3	11
72.2%	27.3%	100%

SURAH 63 MUNAFIQUN: THE HYPOCRITES

Surah 63 Verse Number and Rhyming Letter

1-11									
ن									

Surah 63 Proportional Distribution of Rhyming Letters

ن	TOTAL
11	11
100%	100%

SURAH 64 TAGHABUN: MUTUAL LOSS AND GAIN

Surah 64 Verse Number and Rhyming Letter

1-4	5	6	7-8	9	10	11	12-13	14-15	16
ر	م	د	ر	م	ر	م	ن	م	ن
17-18									
م									

Surah 64 Proportional Distribution of Rhyming Letters

ن	م	ر	د	TOTAL
3	7	7	1	18
16.66%	38.9%	38.9%	5.6%	100%

SURAH 65 TALAQ: DIVORCE

Surah 65 Verse Number and Rhyming Letter

1-5	6	8-12							
ا	*ى	ا							

* pronounced like Alif ا

Surah 65 Proportional Distribution of Rhyming Letters

ا	*ى	TOTAL
11	1	12
91.66%	8.3%	100%

SURAH 66 TAHRIM:HOLDING SOMETHING TO BE FORBIDDEN

Surah 66 Verse Number and Rhyming Letter

1-2	3-4	5	6-7	8-9	10-12				
م	ر	ا	ن	ر	ن				

Surah 66 Proportional Distribution of Rhyming Letters

ن	م	ر	ا	TOTAL
5	2	4	1	12
41.66%	16.66%	33.33%	8.33%	100%

SURAH 67 MULK:DOMINION

Surah 67 Verse Number and Rhyming Letter

1-21	22	23-27	28	29-30					
ر	م	ن	م	ن					

Surah 67 Proportional Distribution of Rhyming Letters

ن	م	ر	TOTAL
7	2	21	30
23.33%	6.66%	70%	100%

SURAH 68 QALAM:PEN

Surah 68 Verse Number and Rhyming Letter

1-3	4	5-10	11-13	14-15	16	17-19	20	21-33	34
ن	م	ن	م	ن	م	ن	م	ن	م

35-39	40	41-47	48-49	50-52					
ن	م	ن	م	ن					

Surah 68 Proportional Distribution of Rhyming Letters

ن	م	TOTAL
42	10	52
88.8%	19.2%	100%

SURAH 69 HAQQAH:THE SURE REALITY
Surah 69 Verse Number and Rhyming Letter

1-18	19-20	21-24	25-26	27	28-32	33	34	35	36-39
ة	ه	ة	ه	ة	ه	م	ن	م	ن

40	41-43	44	45-51	52
م	ن	ل	ن	م

Surah 69 Proportional Distribution of Rhyming Letters

ن	م	ة	ه	ل	TOTAL
15	4	23	9	1	52
28.8%	7.7%	44.2%	17.3%	1.9%	100%

SURAH 70 MA'ARIG:THE WAYS OF ASCENT
Surah 70 Verse Number and Rhyming Letter

1-2	3	4	5-7	8	9	10	11-14	15-18	19-21
ع	ج	ة	ا	ل	ن	ا	ه	ي*	ا

22-23	24-25	26-37	38	39-44
ن	م	ن	م	ن

* pronounced like Alif ا

Surah 70 Proportional Distribution of Rhyming Letters

ن	م	ا	ع	ج	ة	ل	ي*	ه	TOTAL
21	3	7	2	1	1	1	4	4	44
47.7%	6.8%	15.6%	4.5%	2.3%	2.3%	2.3%	9.1%	9.1%	100%

* pronounced like Alif ا

SURAH 71 NUH:NOAH
Surah 71 Verse Number and Rhyming Letter

1	2-4	5-28
م	ن	ا

Surah 71 Proportional Distribution of Rhyming Letters

ن	م	ا	TOTAL
3	1	24	28
10.7%	3.6%	85.7%	100%

SURAH 72 JINN
Surah 72 Verse Number and Rhyming Letter

1-28									
ا									

Surah 72 Proportional Distribution of Rhyming Letters

ا	TOTAL
28	28
100%	100%

SURAH 73 MUSSAMMIL: FOLDED GARMENTS
Surah 73 Verse Number and Rhyming Letter

1	2-19	20							
ل	ا	م							

Surah 73 Proportional Distribution of Rhyming Letters

ل	ا	م	TOTAL
1	18	1	20
5%	90%	5%	100%

SURAH 74 MUDDATHTHIR: ONE WRAPPED UP
Surah 74 Verse Number and Rhyming Letter

1-10	11-14	15	16-17	18-37	38	39-41	42	43-49	50-54
ر	ا	د	ا	ر	ة	ن	ر	ن	ة
55	56								
ه	ة								

Surah 74 Proportional Distribution of Rhyming Letters

ن	ر	ا	ة	ه	د	TOTAL
10	31	6	7	1	1	56
17.9%	55.4%	10.7%	12.5%	1.8%	1.8%	100%

SURAH 75 QIYAMAT: THE RESURRECTION
Surah 75 Verse Number and Rhyming Letter

1-2	3-5	6	7-13	14	15-19	20-25	26	27-30	31-40
ة	ه	ة	ر	ة	ه	ة	ى	قَ	*ى

* pronounced like Alif ا

Surah 75 Proportional Distribution of Rhyming Letters

ة	ه	ر	ى	*ى	قَ		TOTAL
10	8	7	1	10	4		40
25%	20%	17.5%	2.5%	25%	10%		100%

* pronounced like Alif ا

SURAH 76 INSAN: MAN
Surah 76 Verse Number and Rhyming Letter

1-31								
ا								

Surah 76 Proportional Distribution of Rhyming Letters

ا	TOTAL
31	31
100%	100%

SURAH 77 MURSALAT: THOSE SENT FORTH
Surah 77 Verse Number and Rhyming Letter

1-6	7	8-12	13-14	15-21	22	23-24	25-27	28-29	30-31
ا	ع	ت	ل	ن	م	ن	ا	ن	ب
32-33	34-50								
ر	ن								

Surah 77 Proportional Distribution of Rhyming Letters

ن	ع	ب	ر	ا	ت	ل	م		TOTAL
28	1	2	2	9	5	2	1		50
56%	2%	4%	4%	18%	10%	4%	2%		100%

SURAH 78 NABA':THE GREAT NEWS
Surah 78 Verse Number and Rhyming Letter

1	2	3-5	6-40						
ن	م	ن	ا						

Surah 78 Proportional Distribution of Rhyming Letters

ن	م	ا	TOTAL
4	1	35	40
10%	2.5%	87.5%	100%

SURAH 79 NAZI'AT:THOSE WHO TEAR OUT
Surah 79 Verse Number and Rhyming Letter

1-5	6-14	15-26	27-32	33	34-37	38	39-41	42-46	
ا	ة	*ي	ا	م	*ي	ا	*ي	ا	

* pronounced like Alif ا

Surah 79 Proportional Distribution of Rhyming Letters

ا	ة	*ي	م	TOTAL
17	9	19	1	46
37%	19.6%	41.3%	2.2%	100%

* pronounced like Alif ا

SURAH 80 'ABASA:HE FROWNED
Surah 80 Verse Number and Rhyming Letter

1-10	11	12	13-16	17	18	19-24	25-31	32	33
*ي	ة	o	ة	o	o	o	ا	م	ة
34-37	38-42								
o	ة								

* pronounced like Alif ا

Surah 80 Proportional Distribution of Rhyming Letters

*ي	o	ا	ة	م	TOTAL
10	13	7	11	1	42
23.8%	31%	16.7%	26.2%	2.4%	100%

* pronounced like Alif ا

SURAH 81 TAKWIR: THE FOLDING UP
Surah 81 Verse Number and Rhyming Letter

1-14	15-18	19	20-24	25	26-27	28	29		
ت	س	م	ن	م	ن	م	ن		

Surah 81 Proportional Distribution of Rhyming Letters

ن	س	م	ت	TOTAL
8	4	3	14	29
27.6%	13.8%	10.3%	48.3%	100%

SURAH 82 INFITAR: THE CLEAVING ASSUNDER
Surah 82 Verse Number and Rhyming Letter

1-5	6	7-8	9-12	13-14	15-18	19			
ت	م	ك	ن	م	ن	ه			

Surah 82 Proportional Distribution of Rhyming Letters

ن	م	ت	ك	ه	TOTAL
8	3	5	2	1	19
42.1%	15.8	26.3%	10.5%	5.3%	100%

SURAH 83 TATFIR: DEALING IN FRAUD
Surah 83 Verse Number and Rhyming Letter

1-4	5	6-8	9	10-11	12	13-15	16	17-19	20
ن	م	ن	م	ن	م	ن	م	ن	م
21	22	23	24-25	26	27	28-36			
ن	م	ن	م	ن	م	ن			

Surah 83 Proportional Distribution of Rhyming Letters

ن	م	TOTAL
27	9	36
75%	25%	100%

SURAH 84 INSHIQAQ: THE RENDING ASSUNDER

Surah 84 Verse Number and Rhyming Letter

1-5	6-7	8-9	10	11-13	14	15	16-19	20-23	24
ت	ه	ا	ه	ا	ر	ا	قَ	ن	م
25									
ن									

Surah 84 Proportional Distribution of Rhyming Letters

ن	م	ه	ا	ت	قَ	ر	TOTAL
5	1	3	6	5	4	1	25
25%	4%	12%	24%	20%	16%	4%	100%

SURAH 85 BURUJ: THE ZODIACAL SIGNS

Surah 85 Verse Number and Rhyming Letter

1	2-9	10	11	12-18	19	20	21	22	
ج	د	قَ	ر	د	ب	ط	د	ظ	

Surah 85 Proportional Distribution of Rhyming Letters

ج	د	قَ	ر	ب	ط	ظ	TOTAL
1	16	1	1	1	1	1	22
4.5%	72.7%	4.5%	4.5%	4.5%	4.5%	4.5%	100%

SURAH 86 TARIQ: THE NIGHT-VISITANT

Surah 86 Verse Number and Rhyming Letter

1-2	3	4	5-6	7	8-10	11-12	13-14	15-17
قَ	ب	ظ	قَ	ب	ر	ع	ل	ا

Surah 86 Proportional Distribution of Rhyming Letters

قَ	ب	ظ	ر	ع	ل	ا	TOTAL
4	2	1	3	2	2	3	17
23.5%	11.8%	5.9%	17.6%	11.8%	11.8%	17.6%	100%

SURAH 87 A'LA: THE MOST HIGH

Surah 87 Verse Number and Rhyming Letter

1-19								
ي*								

* pronounced like Alif ا

Surah 87 Proportional Distribution of Rhyming Letters

ي*	TOTAL
19	19
100%	100%

* pronounced like Alif ا

SURAH 88 GHASHIYAH: THE OVERWHELMING EVENT

Sura 88 Verse Number and Rhyming Letter

1-5	6-7	8-16	17-20	21-24	25-26			
ة	ع	ة	ت	ر	م			

Surah 88 Proportional Distribution of Rhyming Letters

ة	ع	ت	ر	م	TOTAL
14	2	4	4	2	26
53.8%	7.7%	15.4%	15.4%	7.7%	100%

SURAH 89 FAJR: THE BREAK OF DAY

Surah 89 Verse Number and Rhyming Letter

1-5	6-12	13	14	15-16	17	18	19-22	23	24
ر	د	ب	د	ن	م	ن	ا	ي*	ي
25-26	27-28	29-30							
د	ة	ي							

* pronounced like Alif ا

Surah 89 Proportional Distribution of Rhyming Letters

ن	ر	د	ب	م	ا	ي	ي*	ة	TOTAL
3	5	10	1	1	4	1	3	2	30
10%	16.66%	33.33%	3.33%	3.33%	13.33%	3.33%	10%	6.66%	100%

* pronounced like Alif ا

SURAH 90 BALAD: THE CITY
Surah 90 Verse Number and Rhyming Letter

1-5	6	7	8-10	11-20					
د	ا	د	ن	ة					

Surah 90 Proportional Distribution of Rhyming Letters

د	ا	ن	ة	TOTAL
6	1	3	10	20
30%	5%	15%	50%	100%

SURAH 91 SHAMS: THE SUN
Surah 91 Verse Number and Rhyming Letter

1-15									
ا									

Surah Proportional Distribution of Rhyming Letters

ا	TOTAL
15	15
100%	100%

SURAH 92 LAIL: THE NIGHT
Surah 92 Verse Number and Rhyming Letter

1-21									
ي*									

* pronounced like Alif ا

Surah 92 Proportional Distribution of Rhyming Letters

ي*	TOTAL
21	21
100%	100%

* pronounced like Alif ا

SURAH 93 DHUHA: THE GLORIUS MORNING LIGHT
Surah 93 Verse Number and Rhyming Letter

1-8	9-10	11						
*ي	ر	ث						

* pronounced like Alif ا

Surah Proportional Distribution of Rhyming Letters

*ي	ر	ث	TOTAL
8	2	1	11
72.7%	18.2%	9%	100%

* pronounced like Alif ا

SURAH 94 INSHIRAH: THE EXPANSION
Surah 94 Verse Number and Rhyming Letter

1-4	5-6	7-8						
ك	ا	ب						

Surah 94 Proportional Distribution of Rhyming Letters

ك	ا	ب	TOTAL
4	2	2	8
50%	25%	25%	100%

SURAH 95 TIN: THE FIG
Surah 95 Verse Number and Rhyming Letter

1-3	4	5-8						
ن	م	ن						

Surah 95 Proportional Distribution of Rhyming Letters

ن	م	TOTAL
7	1	8
87.5%	12.5%	100%

SURAH 96 'ALAQA: THE CLOT
Surah 96 Verse Number and Rhyming Letter

1-2	3-5	6-14	15-16	17	18	19			
قَ	م	ي*	ة	ه	ة	ب			

* pronounced like Alif

Surah 96 Proportional Distribution of Rhyming Letters

م	قَ	ي*	ة	ه	ب	TOTAL
3	2	9	3	1	1	19
15.85	10.5%	47.4%	15.85%	5.3%	5.3%	100%

* pronounced like Alif

SURAH 97 QUADR: THE NIGHT OF POWER
Surah 97 Verse Number and Rhyming Letter

1-5									
ر									

Surah 97 Proportional Distribution of Rhyming Letters

ر	TOTAL
5	5
100%	100%

SURAH 98 BAIYINA: THE CLEAR EVIDENCE
Surah 98 Verse Number and Rhyming Letter

1-7	8								
ة	ه								

Surah 98 Proportional Distribution of Rhyming Letters

ة	ه	TOTAL
7	1	8
87.5%	12.5%	100%

SURAH 99 ZILZAL: THE CONVULSION
Surah 99 Verse Number and Rhyming Letter

1-5	6	7-8							
ا	م	ه							

Surah 99 Proportional Distribution of Rhyming Letters

ا	م	ه	TOTAL
5	1	2	8
62.5%	12.5%	25%	100%

SURAH 100 ADIYAT: THOSE THAT RUN

Surah 100 Verse Number and Rhyming Letter

1-5	6-8	9-11						
ا	د	ر						

Surah 100 Proportional Distribution of Rhyming Letters

ا	د	ر	TOTAL
5	3	3	11
45.5%	27.3%	27.3%	100%

SURAH 101 AL-QARI'AH: THE DAY OF NOISE AND CLAMOUR

Surah 101 Verse Number and Rhyming Letter

1-3	4	5	6	7	8	9	10	11
ة	ث	ش	ه	ة	ه	ة	ه	ة

Surah 101 Proportional Distribution of Rhyming Letters

ة	ث	ش	ه	TOTAL
6	1	1	3	11
54.5%	9%	9%	27.3%	100%

SURAH 102 TAKATHUR: PILING UP

Surah 102 Verse Number and Rhyming Letter

1-2	3-5	6	7	8				
ر	ن	م	ن	م				

Surah 102 Proportional Distribution of Rhyming Letters

ن	م	ر	TOTAL
4	2	2	8
50%	25%	25%	100%

SURAH 103 'ASR: TIME THROUGH THE AGES
Surah 103 Verse Number and Rhyming Letter

1-3									
ر									

Surah 103 Proportional Distribution of Rhyming Letters

ر	TOTAL
3	3
100%	100%

SURAH 104 HUMAZA: THE SCANDEL-MONGER
Surah 104 Verse Number and Rhyming Letter

1	2-3	4-9							
ة	ه	ة							

Surah 104 Proportional Distribution of Rhyming Letters

ة	ه	TOTAL
7	2	9
77.8%	22.2%	100%

SURAH 105 FIL: THE ELEPHANT
Surah 105 Verse Number and Rhyming Letter

1-5									
ل									

Surah 105 Proportional Distribution of Rhyming Letters

ل	TOTAL
5	5
100%	100%

SURAH 106 QUARAISH
Surah 106 Verse Number and Rhyming Letter

1	2	3	4						
ش	ف	ت	ف						

Science Miracles: No Stick or Snakes

Surah 106 Proportional Distribution of Rhyming Letters

ش	ت	ف	TOTAL
1	1	2	4
25%	25%	50%	100%

SURAH 107 MAUN:NEIGHBOURLY NEEDS

Surah 107 Verse Number and Rhyming Letter

1	2	3-7							
ن	م	ن							

Surah 107 Proportional Distribution of Rhyming Letters

ن	م	TOTAL
6	1	7
85.7%	14.3%	100%

SURAH 108 KAUTHAR:ABUNDANCE

Surah 108 Verse Number and Rhyming Letter

1-3									
ر									

Surah 108 Proportional Distribution of Rhyming Letters

ر	TOTAL
3	3
100%	100%

SURAH 109 KAFIRUN:THOSE WHO REJECT FAITH

Surah 109 Verse Number and Rhyming Letter

1-2	3	4	5	6					
ن	د	م	د	ن					

Surah 109 Proportional Distribution of Rhyming Letters

ن	د	م	TOTAL
3	2	1	6
50%	33.33%	16.66%	100%

SURAH 110 NASR:HELP
Surah 110 Verse Number and Rhyming Letter

1	2-3							
ح	ا							

Surah 110 Proportional Distribution of Rhyming Letters

ا	ح	TOTAL
2	1	3
66.66%	33.33%	100%

SURAH 111 LAHAB: THE FATHER OF FLAME
Surah 111 Verse Number and Rhyming Letter

1-4	5							
ب	د							

Surah 111 Proportional Distribution of Rhyming Letters

ب	د	TOTAL
4	1	5
80%	20%	100%

SURAH 112 IKHLAS: PURITY OF FAITH
Surah 112 Verse Number and Rhyming Letter

1-4								
د								

Surah 112 Proportional Distribution of Rhyming Letters

د	TOTAL
4	4
100%	100%

SURAH 113 FALAQ: THE DAWN
Surah 113 Verse Number and Rhyming Letter

1-2	3	4-5						
ق	ب	د						

Surah 113 Proportional Distribution of Rhyming Letters

قَ	بِ	دِ	TOTAL
2	1	2	5
40%	20%	40%	100%

SURAH 114 NAS:MANKIND
Surah 114 Verse Number and Rhyming Letter

1-6								
س								

Surah 114 Proportional Distribution of Rhyming Letters

س	TOTAL
6	6
100%	100%

Master Table of Rhyming Letters in the Qur'an.

surah No	ا	ـ	د	د	ع	ح	ـ	ذ	ن	ب	ك	ل	م	ـ	ـ	ز	ن	ل	ح	ر	ه	٥	ن	ي	ى	ء	Total
1																											7
2	9			7		21													1	3	54	193					286
3	9			9		23													3	30	121						200
4	170																		1	5							176
5	4			2		7													3	24	80						120
6						4													3	13	144						165
7								1			1								2	10	193				3		206
8	4			1		10								1					1	19	39						75
9	1					4													1	37	86						129
10						11													1	10	98						109
11	11			23	3	11		1		4		3							2	5	56						123
12				1		1													1	15	93						111
13				4		8					3								7		5						43
14	15			10		14	1												4	7	6				4		52
15						2													2	16	81						99
16						1														16	110						128
17	110																										111
18	109							1																			110
19	90																			2	5						98
20	26																			1				19	88		135
21																				6	106						112
22	2			15		25	2							1	1				1	12	12					1	78
23																		6		4	114						118
24	2					7													1	23	31						31
25	76																										77
26																			1	31	192						227
27																			4	9	84						93

Master Table of Rhyming Letters in the Qur'ān (continued).

surah No	ا	ب	ت	ث	ج	ح	خ	د	ذ	ر	ز	س	ش	ص	ض	ط	ظ	ع	غ	ف	ق	ك	ل	م	ن	ه	و	ي	ى	.	Total
28																															88
29					2			3														2		3	7	99					69
30						2																			4	54					60
31						2			16	1															8	7					34
32	72																								2	27					73
33							5																	1	4	22					30
34	7					1	9		12	1		1		1						1					8	1					54
35	3					3			29																4						45
36						5	2			1		2														145					83
37							6		15			2							1				6		26						182
38						35			15														1	3	18						88
39					4		6																	2	5	53			1			75
40			17		10																			3	5	32						85
41			1		7				2	1							1							8	30						54
42			5		4				20	1		1				1								4	11	6					53
43										1														1	10	78					89
44																								15	44					59
45		1								1										1			7	30						37
46							1														1		8	26						35
47	2																						36							38
48	29							1															11							29
49						1																	7	10						18
50	7		5	27		2				2					1									9	42						45
51	4								1													2		8	26						60
52	3										5														30	3					49
53	4																												53		62

Master Table of Rhyming Letters in the Qur'an (continued).

surah No	ا	ـ	د	ذ	ر	ز	ج	ح	خ	س	ش	ص	ض	ط	ظ	ع	غ	ف	ق	ك	ل	م	ن	ه	و	ى	ء	Total
54											55																	55
55											2										8	68						78
56	7	1			3															1	18	55	10				1	96
57					1			1			12	1									10	5						29
58					1						5	1									3	12						22
59		2									3									1	5	14						24
60					1						3										4	4						13
61																					3	10						14
62										1											3	8						11
63																						11						11
64					1						7										7	3				1		18
65	11																					3						12
66	1										4									1	2	5						12
67											21										2	7						30
68																					10	43						52
69																1					4	15	9	23				52
70	7				1																3	21	4	1		4		44
71	24																				1	3						28
72	28																											28
73	18																			1	1							20
74	6										31											10	1	7				56
75	31										7												8	10	1	10		40
76																												31
77	9	2																		2	1	28						50
78	35		5								2										1	4						40
79	17																				1			9		19		46

Master Table of Rhyming Letters in the Qur'an (continued).

surah No	ا	ـا	ـا	ع	ح	د	ذ	ر	ز	س	ش	ص	ض	ط	ظ	ع	غ	ف	ق	ك	ل	م	ن	ه	ة	ي	ى	·	Total
84	6																		4			1	5	3					25
85		1	5								3																		23
86	3	2						16		1									1										22
87																											19		19
88			4						4					2				4				2			14				26
89	4	1						10	5													1	3		2	1	3		30
90	1							6															3		10				20
91	15		1																										15
92									2																		21		21
93		1																									8		11
94	2	2							2							4													8
95																					5		1						15
96	1								3								2					3	1		3		9		19
100	5			3				3	3																				11
101					1																		1						11
102				1					2													2		3	6				8
103									3													4							3
104																								2	7				9
105																					5								5
106																2													4
107			1																			1							3
108								2		3														3					7
109																								6					6
110	2					1																							3

Master Table of Rhyming Letters in the Qur'an (continued).

surah No	ا	ب	ت	ث	ج	ح	خ	د	ذ	ر	ز	س	ش	ص	ض	ط	ظ	ع	غ	ف	ق	ك	ل	م	ن	ه	و	ي	ى	ء	Total
1																															
111	4							1		4																					5
112																							2								4
113		1							2																						5
114															6																6
Total	949	160	34	2	9	1	199	3	453	9	11	2	10	1	12	13	13	4	40	8	67	666	3123	49	122	22	246	9			6236
Total %	15.32	2.57	0.55	0.33	0.14	.02	3.18	.05	7.26	.14	.18	.03	.16	.02	.19	.21	.21	.06	.64	.13	1.07	10.68	50.08	.79	1.96	.35	3.94	.14			100

Table 12: Showing surahs with verses rhyming with alif and yā with a maddah which is pronounced like alif.

Surah	Alif %	Yā %	Surah	Alif %	Yā %
4: Nisā	96.6	0	75:Qiyamat	0	25
17:Al-Isrā	99.1	0	76:Insān	100	0
18:Kahf	99.1	0.9	77:Mursalat	18	0
19:Maryam	91.8	0	78:Naba	87.5	0
20:Taha	19.3	65.2	79:Nazi'at	37	41.3
25:Furqan	98.7	0	80:'Abasa	16.7	23.8
33:Aḥzab	98.6	0	84:Inshiqaq	24	0
37:Saffat	1.6	0	86:Ṭariq	17.6	0
47:Muhammad	5.3	0	87:A'la	0	100
48:Fatḥa	100	0	89:Fajr	13.33	10
52:Ṭūr	6.1	0	90:Balad	5	0
53:Najm	6.5	85.5	91:Shams	100	0
65:Ṭalaq	91.6	8.3	92:Layl	0	100
66:Taḥrim	8.33	0	93:Dhuha	0	72.7
70:Ma'arij	15.6	9.1	94:Inshirah	25	0
71:Nuḥ	85.7	0	96:'Alaq	0	47.4
72:Jinn	100	0	99:Zilzal	62.5	0
73:Muzzamil	90	0	100:Adiyat	45.5	0
74:Muddaththir	10.7	0	110:Nasr	66.66	0

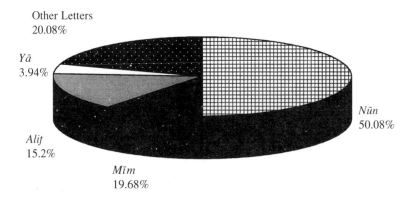

Figure 23: Pie graph to show proportionate distribution of rhyming letters in the Qur'an.

Science Miracles: No Stick or Snakes

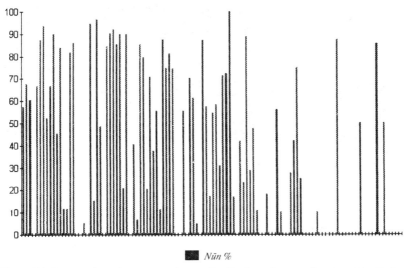

Figure 24: *Bar graph to show proportionate distribution of verses rhyming with* nūn *in the* surahs *of the Qur'an.*

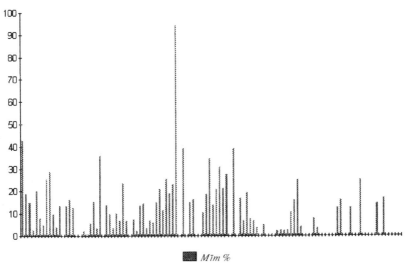

Figure 25: *Bar graph to show proportionate distribution of verses rhyming with* mīm *in the* surahs *of the Qur'an. NB: The highest spike is in* surah *47, Muhammad, perhaps in his honor.*

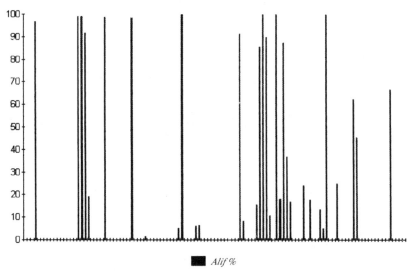

Figure 26: Bar graph to show proportionate distribution of verses rhyming with alif in the surahs of the Qur'an.

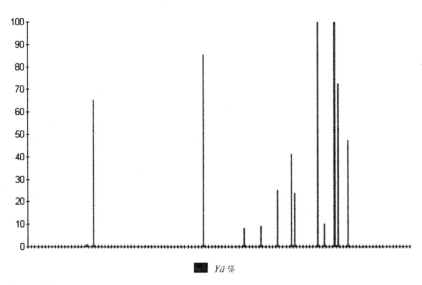

Figure 27: Bar graph to show proportionate distribution of verses rhyming with yā in the surahs of the Qur'an.